COMPLEX
THERMODYNAMIC
SYSTEMS

STUDIES IN SOVIET SCIENCE

1973

STUDIES IN SOVIET SCIENCE

COMPLEX THERMODYNAMIC SYSTEMS

V. V. Sychev

Moscow Physicotechnical Institute
Moscow, USSR

Translated from Russian by

John S. Shier

Signetics Memory Systems
Sunnyvale, California

CONSULTANTS BUREAU • NEW YORK-LONDON

Library of Congress Cataloging in Publication Data

Sychev, Viacheslav Vladimirovich.
  Complex thermodynamic systems.

  (Studies in Soviet science)
  Translation of Slozhnye termodinamicheskie sistemy.
  Bibliography: p.
  1. Thermodynamics. I. Title. II. Series.
QC311.S92513   536'.7   73-83896
ISBN 978-1-4684-1607-7   ISBN 978-1-4684-1605-3 (eBook)
DOI 10.1007/978-1-4684-1605-3

Vyacheslav Vladimirovich Sychev was born in 1933 and was graduated from the Moscow Institute of Power Engineering in 1957. He is Deputy Director of the Institute of High Temperatures of the Academy of Sciences of the USSR and a professor at the Moscow Physicotechnical Institute. Professor Sychev is noted in the Soviet Union as an expert in the thermodynamic properties of materials.

The original Russian text, published by Energiya Press in Moscow in 1970, has been corrected by the author for the present edition. This translation is published under an agreement with Mezhdunarodnaya Kniga, the Soviet book export agency.

SLOZHNYE TERMODINAMICHESKIE SISTEMY
V. V. Sychev
Сложные термодинамические системы
В. В.-Сычев

© 1973 Consultants Bureau, New York
Softcover reprint of the hardcover 1st edition 1973
A Division of Plenum Publishing Corporation
227 West 17th Street, New York, N.Y. 10011

United Kingdom edition published by Consultants Bureau, London
A Division of Plenum Publishing Company, Ltd.
Davis House (4th Floor), 8 Scrubs Lane, Harlesden, London, NW10 6SE, England

Preface to the American Edition

Thermodynamic methods of analysis have in recent years found ever-growing extensions in diverse regions of modern technology. The object of the present book is to apply these methods to the description of materials of varying physical properties. I hope the book will illustrate the wide variety and usefulness of thermodynamics which was well described by Albert Einstein: "A theory is the more impressive the greater the simplicity of its premises is, the more different kinds of things it relates, and the more extended is its area of applicability. Therefore the deep impression which classical thermodynamics made upon me."

The work of the American thermodynamic school is well known in the Soviet Union, and thus it is a great pleasure to offer this book to American readers.

V. V. Sychev

Preface

At the present time, when a number of new areas of technology are rapidly evolving, it is difficult to present a modern course in technical thermodynamics without developing such subjects as the thermodynamics of insulators, magnets, and superconductors, or without treating the features of thermodynamic systems located in a gravitational field and in conditions of weightlessness, etc. In fact the limited coverage of technical thermodynamics in the usual textbooks and school equipment as a rule prevents the authors from giving any detailed discussion of these important problems. I therefore resolved to treat these problems in a separate text.

I discussed the concept of this book with my teachers V. A. Kirillin and A. E. Sheindlin when we were working together on our textbook "Technical Thermodynamics," which was published in 1968. The present book can be thought of as a supplement to that book in its methodological structure and the exposition scheme used. The aim of the book is to develop, from a unified point of view, a description of the various physical phenomena and processes which are of primary interest for modern technology and in particular to demonstrate the universality of thermodynamic methods applied to the solution of various problems. It was felt desirable to develop the technical aspects of the problems more prominently.

The title of this book is undoubtedly a poor choice, but it more or less reflects the contents — "technical thermodynamics of systems which, in addition to work of expansion, involve other forms of work." Such a title would be too cumbersome to use.

The book is primarily directed toward students in thermo-physical, physicotechnical, and engineering physics areas; the book is based on material from a course of lectures given by the author to students at the Moscow Physicotechnical Institute. It is hoped that in some respects the book will also be useful to scientific workers specializing in the areas covered.

It is assumed that the reader has had the usual course in general or technical thermodynamics. Background information of a handbook character is briefly developed in Chapter 1 ("Introduction").

I would like to express my gratitude to all whose advice, remarks, and criticism assisted in the work on this book. I am particularly grateful to V. A. Kirillin and A. E. Sheindlin for the interest and attention given to this work, and to L. I. Gordon, L. M. Biberman, M. G. Kremlev, and the editor of the book, S. P. Malyshenko, for much valuable advice. V. B. Zenkevich has been of great assistance in the work and provided the material in Chapter 8.

Any critical remarks from readers would be appreciated.

V. V. Sychev

Contents

CHAPTER 1

Introduction

1.1. The Equations of the First and Second Laws of Thermodynamics

The well-known equation for the first law of thermodynamics — the conservation and conversion of energy — can be written in differential form

$$dQ = dU + dL, \tag{1.1}$$

where Q is the amount of heat supplied to a thermodynamic system (or given up by it), U is the internal energy of the system, and L is the work done by the system. This equation states that heat supplied to a system in general increases its internal energy or goes into the performance of work.

This equation has the integral form

$$Q_{1-2} = (U_2 - U_1) + L_{1-2}. \tag{1.2}$$

Here the subscripts 1 and 2 refer to the initial and final states of the system in a certain 1-2 thermodynamic process.

We recall that Q and L are not state functions, but depend on the process by which the system goes from state 1 to state 2. In other words, the differentials dQ and dL are not total differentials (this will be demonstrated in Section 1.5).

The equation for the second law of thermodynamics takes the form

1

$$TdS \geqslant dQ, \tag{1.3}$$

where S is the entropy of the system. Here the "greater than" relation is valid when the system undergoes an irreversible process while the equality is valid for a reversible process. Thus for reversible processes

$$dQ = TdS \tag{1.4}$$

or in integral form

$$Q_{1-2} = \int_{1}^{2} T \, dS, \tag{1.5}$$

where Q_{1-2} is the heat supplied to the system (or given up by it) in a reversible process between states 1 and 2.

It is evident from relations (1.1) and (1.3) that a combined equation for the first and second laws of thermodynamics can be written in the form

$$TdS \geqslant dU + dL. \tag{1.6}$$

In view of the remarks above, when the system undergoes a reversible process this relation takes the form

$$TdS = dU + dL, \tag{1.7}$$

and in integral form

$$\int_{1}^{2} TdS = (U_2 - U_1) + L_{1-2}. \tag{1.8}$$

These relations are written for the total values of U, L, and S for the whole system. These equations can also be written for the weight specific values of these quantities

$$Tds = du + dl \tag{1.7a}$$

and

$$\int_{1}^{2} Tds = (u_2 - u_1) + l_{1-2}, \tag{1.8a}$$

where u = U/G, l = L/G, s = S/G, and G is the weight of the material in the system. The relations can be written in a similar way for the volume-specific values of these quantities or for their molar values.

Quantities which determine the state of a system are either intensive or extensive. By intensive quantities we mean those whose values do not depend on the amount of material in the system (for example, pressure and temperature) while the extensive quantities do depend on the amount of material in the system (for example, the volume). The extensive quantities are additive. The specific extensive quantities (i.e., the values per unit amount of material) behave like intensive quantities (for example, the specific volume and the specific heat capacity are intensive quantities).

In what follows we will need to use the equation for the first law of thermodynamics under flow conditions. We recall that this equation can be written in differential form for the flowing liquid or gas moving in a channel using weight-specific thermodynamic quantities as

$$dq = di + \frac{w\,dw}{g} + dh + dl_{\text{tech}} + dl_{\text{d}}, \qquad (1.9)$$

and in integral form

$$q_{1-2} = (i_2 - i_1) + \left(\frac{w_2^2 - w_1^2}{2g} \right) + (h_2 - h_1) + l_{\text{tech}} + l_{\text{d}}, \qquad (1.10)$$

where q is the heat supplied to the flowing material (or given up by it), i is the enthalpy of the liquid or gas, w is the flow velocity, h is the height, l_{tech} is the so-called technical flow work,[†] and l_{d} is the dissipative work (for example, the work done by the flow in in overcoming frictional forces).

[†]One example of technical work done by a flow is rotation of a turbine wheel. If the flow involves an electrically conductive liquid in a transverse magnetic field the technical work involves transfer of electrical energy to an external circuit due to magnetohydrodynamic effects, etc. Not only can technical work be done by the flow but it can be done on the flow (here one can cite examples of the opposite kind: the flow can be forced by a centrifugal pump, induced by an electromagnetic pump, etc.).

The heat q which enters Eqs. (1.9) and (1.10) can be divided into two parts — the heat q_{ext} brought into the flowing material from outside (or given up by it to the external medium) and the dissipative heat q_d which appears, for example, when there is friction in the flow:

$$q = q_{ext} + q_{d}.$$ (1.11)

Since the dissipative heat q_d is equivalent to work l_d, Eqs. (1.9) and (1.10) can be rewritten in the form

$$dq_{ext} = di + \frac{wdw}{g} + dh + dl_{tech}$$ (1.12)

and

$$q_{ext\ _{1-2}} = (i_2 - i_1) + \left(\frac{w_2^2 - w_1^2}{2g}\right) + (h_2 - h_1) + l_{tech},$$ (1.13)

it being understood that these equations are valid with and without friction in the flow.

1.2. Work

It is known from mechanics that the work L_{1-2} for a certain arbitrary force $\mathfrak{F}$ is given by the integral

$$L_{1-2} = \int_1^2 \widetilde{\mathfrak{F}} dx,$$ (1.14)

where $\widetilde{\mathfrak{F}}$ is the projection of the force $\mathfrak{F}$ along the direction of the element of the body's displacement (or to be more precise, the displacement of the coordinates giving the body's position under the action of the force, which we denote by dx.

One particular case of this equation is the well-known equation for the work of expansion of a thermodynamic system against external pressure forces. We recall that this equation is found as follows. We consider an increase in the volume V of a body of arbitrary shape located in a medium at a pressure p_c (see Fig. 1.1). The surface area of the body is denoted by $\mathfrak{S}$. If the change in the volume of the body is regarded as infinitesimal (dV) then the increase in volume can be regarded as a translation of each point on the surface of this body by a distance dx. Since the pressure is the force acting along the normal per unit surface area of the

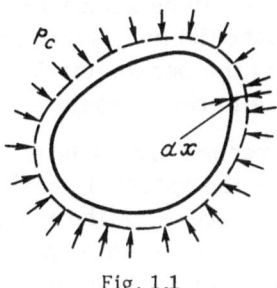

Fig. 1.1

body it is clear that the sum of the forces acting on the whole surface of the body is $\tilde{\mathfrak{F}} = p_c \mathfrak{S}$.

It is clear from this that the work which must be expended
to move the surface $\mathfrak{S}$ by a distance dx against the external pressure is

$$dL = p_c \mathfrak{S} dx, \qquad (1.15)$$

and since evidently

$$\mathfrak{S} dx = dV, \qquad (1.16)$$

then

$$dL = p_c \, dV \qquad (1.17)$$

and

$$L_{1-2} = \int_{V_1}^{V_2} p_c dV. \qquad (1.18)$$

This is the relation for the expansion work performed by the system on the surrounding medium.

It should be stressed that expansion work against external
pressure forces only occurs when the volume V changes and external bodies are displaced. If V remains constant, then despite
changes in other parameters characterizing the states of the body
(temperature, internal energy, potential energy in a gravitational
field, etc.), the expansion work is zero. On the other hand, the
work done by a gas in expanding in vacuum is zero despite the

fact that V changes. This can be seen from (1.18) since $p_c = 0$. Thus if we inquire whether the body (system) does work against the force p_c, we see that the parameter V is related to this force (or, as we say, conjugate to this force).

If the expansion process proceeds in equilibrium then at each point in the process the external pressure acting inward on the system (i.e., the pressure p_c of the medium) will equal the pressure within the system (denoted by p). Since we will only consider equilibrium processes except when otherwise noted, we will use the relation

$$dL = pdV \qquad (1.17a)$$

for the work, and

$$L_{1-2} = \int_{V_1}^{V_2} pdV, \qquad (1.18a)$$

obtained from (1.17) and (1.18) by replacing p_c by p.

It is not hard to see that Eq. (1.18a) is a particular case of Eq. (1.14). Other forms of work can be performed by a system besides work of expansion, for example, work due to an increase in the surface area against surface tension forces, work due to motion in a gravitational, electric, or magnetic field, etc. The analysis of such systems is the object of the present book. As we shall see, despite the great differences in all of these forms of work, there is a common relation for calculating the work in all cases which is structurally identical and analogous to Eq. (1.14):

$$dL = \zeta \, dY \qquad (1.19)$$

and

$$L_{1-2} = \int_1^2 \zeta dY. \qquad (1.20)$$

Here ζ is the external force acting on the body (system) while Y is the state parameter (coordinate) of the system which is con-

jugate to the force ζ. We refer to ζ as a generalized force
and Y as a generalized coordinate.†

We again stress that different forms of generalized force
will have different generalized coordinates conjugate to them.
Everywhere below in considering particular systems we will always
establish which state parameters of the given system are generalized forces and which are generalized coordinates.

It is not difficult to see a definite analogy between Eqs. (1.19) and (1.20) on
the one hand and the relations for calculating the amount of heat

$$dQ = TdS \tag{1.4}$$

and

$$Q_{1-2} = \int_1^2 TdS. \tag{1.5}$$

In considering the amount of heat the temperature plays the role of a generalized force
and the entropy the role of a generalized coordinate.

If several different kinds of force act simultaneously in a
system then evidently the work done by the system will be the sum
of the amounts of work done by the system due to each force

$$dL = \sum_{i=1}^{n} \zeta_i dY_i \tag{1.21}$$

and

$$L_{1-2} = \sum_{i=1}^{n} \int_1^2 \zeta_i dY_i, \tag{1.22}$$

where ζ_i is the i-th generalized force, Y_i is the generalized coordinate conjugate to the i-th force, and n is the number of generalized
forces.

It should be understood that for the weight-specific values
of the thermodynamic quantities which enter Eqs. (1.17a), (1.18a),

---

†Sometimes a generalized force is called an intensity factor and a generalized
coordinate is called a capacity factor. These names indicate that generalized
forces are intensive quantities while generalized coordinates are extensive quantities.

(1.19), and (1.22), these equations are written as

$$dl = pdv, \tag{1.17b}$$

$$l_{1-2} = \int_{v_1}^{v_2} pdv, \tag{1.18b}$$

$$dl = \zeta dy, \tag{1.19a}$$

$$l_{1-2} = \int_{1}^{2} \zeta dy, \tag{1.20a}$$

$$dl = \sum_{i=1}^{n} \zeta_i dy, \tag{1.21a}$$

$$l_{1-2} = \sum_{i=1}^{n} \zeta_i dy, \tag{1.22a}$$

where v is the specific volume and y are weight-specific values of the generalized coordinates Y (y = Y/G, where G is the weight of material in the system).

In Chapter 2, in studying equilibrium in thermodynamic systems, we will also consider systems which have no more than two different forms of work, one of which is work of expansion. In this regard it is appropriate to represent the work done by the system as two terms — the work of expansion and any other possible form of work. We will use the convention that any form of work in general will be denoted by L, while any form of work with the exception of work of expansion will be denoted by L*.

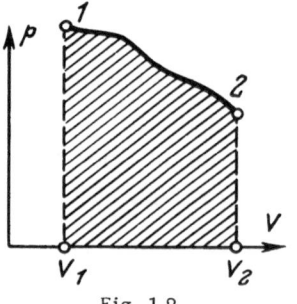

Fig. 1.2

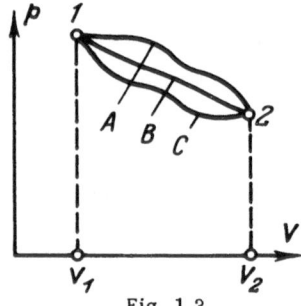

Fig. 1.3

Then in agreement with this notation

$$dL = pdV + dL^*,\tag{1.23}$$

$$L_{1-2} = \int_{V_1}^{V_2} pdV + L^*{}_{1-2},\tag{1.24}$$

while using weight-specific quantities

$$dl = pdv + dl^*,\tag{1.23a}$$

$$l_{1-2} = \int_{v_1}^{v_2} pdv + l^*{}_{1-2}.\tag{1.24a}$$

In addition we will use the notation

$$dL^* = \xi dX,\tag{1.25}$$

$$dl^* = \xi dx,\tag{1.25a}$$

where ξ is any generalized force (with the exception of the pressure p), X is any generalized coordinate (with the exception of the volume V), and x = X/G.

The amount of work done by the system is conveniently calculated using a diagram in which the ordinate gives the generalized force and the abscissa the generalized coordinate. We consider as an illustration the change in the volume of a system undergoing work of expansion in a p-V diagram (Fig. 1.2). The volume of the system goes from V_1 to V_2. The states through which the system passes during the volume change lie on the curve for the process between points 1 and 2. It is evident from Eqs. (1.18a) that the work of expansion for the system is shown by the area under the curve for the process in a p-V diagram (the shaded area in Fig. 1.2).

As previously noted, work and heat are not functions of state but functions of the process. In other words they depend not only on the parameters of the initial and final states for the process of interest but also on the path by which the process occurs. For the work of expansion this is evident in particular from the p-V diagram; we see from Fig. 1.3 that the magnitude of the integral

$$L_{1-2} = \int_{V_1}^{V_2} pdV$$

will differ depending on which path (A, B, C, etc.) the expansion process takes.

Using the concepts and notation introduced in this section, the combined equation for the first and second laws of thermodynamics can be written in the following forms. Substituting dL from (1.21) into (1.6), we find

$$T dS \geqslant dU + \sum_{i=1}^{n} \zeta_i dY_i, \qquad (1.26)$$

or, using (1.23),

$$T dS \geqslant dU + p dV + dL^*, \qquad (1.27)$$

or, what is the same [see (1.25)],

$$T dS \geqslant dU + p dV + \xi dX. \qquad (1.28)$$

When the only form of work undergone by the system is work of expansion this relation is written as

$$T dS \geqslant dU + p dV. \qquad (1.29)$$

For systems in equilibrium

$$T dS = dU + \sum_{i=1}^{n} \zeta_i dY_i, \qquad (1.26a)$$

or

$$T dS = dU + p dV + dL^*, \qquad (1.27a)$$

or

$$T dS = dU + p dV + \xi dX, \qquad (1.28a)$$

and correspondingly for the weight-specific values

$$T ds = du + \sum_{i=1}^{n} \zeta_i dy_i, \qquad (1.26b)$$

$$T ds = du + p dv + dl^*, \qquad (1.27b)$$

$$Tds = du + pdv + \xi dx. \tag{1.28b}$$

In what follows we assume that work done by a system is positive while work done on a system is negative.

1.3. Heat Capacities

The heat capacity is known to be given by the relation

$$c_z = \frac{dq_z}{dT}, \tag{1.30}$$

where c_z is the heat capacity in a process in which some parameter z is kept constant.

Since

$$dq = Tds, \tag{1.31}$$

it follows from (1.30) that

$$c_z = T \left(\frac{\partial s}{\partial T} \right)_z. \tag{1.32}$$

The equation for the isochoric heat capacity c_v is written as follows (x = v = const)

$$c_v = T \left(\frac{\partial s}{\partial T} \right)_v = \left(\frac{\partial u}{\partial T} \right)_v, \tag{1.33}$$

while for the isobaric heat capacity c_p (x = p = const),

$$c_p = T \left(\frac{\partial s}{\partial T} \right)_p = \left(\frac{\partial i}{\partial T} \right)_p. \tag{1.34}$$

Evidently these equations are only valid when the only form of work is work of expansion (i.e., when $dl^* = 0$). When $dl^* \neq 0$, using (1.27b)

$$Tds = du + pdv + dl^*,$$

it follows from (1.33) that the isochoric heat capacity of a system which undergoes other forms of work in addition to work of expansion (we denote this heat capacity by c_v^*) is given by the relation

$$c^*{}_v = c_v + \left(\frac{\partial l^*}{\partial T} \right)_v.$$

(1.35)

In a similar way we find for the isobaric heat capacity of such a system:

$$c^*{}_p = c_p + \left(\frac{\partial l^*}{\partial T} \right)_p.$$

(1.36)

It should be stressed that whenever we speak of the heat capacity of a system which undergoes other kinds of work besides work of expansion we must carefully state which heat capacity is meant — whether or not we take account of $(\partial l^*/\partial T)_z$ — since in many cases the value of this "correction" is quite significant and its neglect can lead to substantial errors.

By analogy with the heat capacities c_p and c_v for systems undergoing only work of expansion, for many systems which will be treated in the following chapters we will be greatly interested in the heat capacities at constant generalized force, c_ξ, and at constant generalized coordinate, c_x. In agreement with the general definition (1.32) these heat capacities will be given by the following relations:

$$c_\xi = T \left(\frac{\partial s}{\partial T} \right)_\xi$$

(1.37)

and

$$c_x = T \left(\frac{\partial s}{\partial T} \right)_x.$$

(1.38)

1.4. The Differential Equations of Thermodynamics

The mathematical apparatus used in thermodynamics is quite simple. We will give several basic relations which will be required in what follows.

As noted above, all thermodynamic quantities are either state functions or process functions. The magnitude of a state function is uniquely determined by the parameters of the given state. Consequently, to determine the changes in a state function during any process we need only know the values of this function at the beginning and end of the process. As examples of state functions we can mention the specific volume, entropy, enthalpy, etc. As for quantities which are process functions, they are characteristic of the process and their values in a given state depend not only on the parameters of this state but also on the path (i.e., the process) by which the system reached the given state. Examples of process functions include the work done by the system in going from one state to another, or the heat. As noted previously (Section 1.1), the characteristic property of a state function is that its differential is a total differential.

As we know, the total differential of a function of several independent variables $z = f(x, y, w \ldots)$ is given by

$$dz = \left(\frac{\partial z}{\partial x}\right)_{y, w \ldots} dx + \left(\frac{\partial z}{\partial y}\right)_{x, w \ldots} dy + \left(\frac{\partial z}{\partial w}\right)_{x, y \ldots} dw + \cdots . \qquad (1.39)$$

In the overwhelming majority of cases below we will consider functions of two variables for which

$$dz = \left(\frac{\partial z}{\partial x}\right)_{y} dx + \left(\frac{\partial z}{\partial y}\right)_{x} dy. \qquad (1.40)$$

The indices attached to a partial derivative show that it is taken holding the quantities in the index constant. For example, the derivative of pressure with respect to temperature, $\partial p/\partial T$, can be calculated under various conditions: at constant volume V, at constant enthalpy I, constant entropy S, etc. In these cases the derivative under consideration is denoted by $(\partial p/\partial T)_V$, $(\partial p/\partial T)_I$, $(\partial p/\partial T)_S$, etc., and will have a different value. We remind the reader that, as we know from mathematical analysis, for a state function $z(x, y)$

$$\frac{\partial^2 z}{\partial x \partial y} = \frac{\partial^2 z}{\partial y \partial x}, \qquad (1.41)$$

i.e., mixed derivatives do not depend on the order of differentiation.

It is evident from this that if the differential of any function $z = f(x, y)$ is written in the form

$$dz = M dx + N dy, \qquad (1.42)$$

and if we know that the function z has the property that its differential is a total differential, then the following relation is valid:

$$\left(\frac{\partial M}{\partial y}\right)_x = \left(\frac{\partial N}{\partial x}\right)_y. \qquad (1.43)$$

This relation will enable us to obtain useful equations.

It is possible, for example, to use this relation to show that the differential of a process function is not a total differential. We consider, for example, an expression for the differential of the amount of heat supplied to the system, i.e., the equation of the first law of thermodynamics when the system undergoes some single form of work. In this case we find from Eq. (1.1), taking (1.19) into account,

$$dQ = dU + \zeta dY. \qquad (1.44)$$

In order to represent Q as a function of two variables ζ and Y we write dU in this equation with the aid of (1.40)

$$dU = \left(\frac{\partial U}{\partial T}\right)_Y dT + \left(\frac{\partial U}{\partial Y}\right)_T dY, \qquad (1.45)$$

and find

$$dQ = \left(\frac{\partial U}{\partial T}\right)_Y dT + \left[\left(\frac{\partial U}{\partial Y}\right)_T + \zeta\right] dY. \qquad (1.46)$$

We now check to see whether Eq. (1.43) is satisfied for this relation. When we apply it to (1.46) we find

$$M = \left(\frac{\partial U}{\partial T}\right)_Y, \quad N = \left(\frac{\partial U}{\partial Y}\right)_T + \zeta, \quad x = T, \; y = Y.$$

from which it follows that

$$\left(\frac{\partial M}{\partial y}\right)_x = \frac{\partial^2 U}{\partial T \, \partial Y},$$

$$\left(\frac{\partial N}{\partial x}\right)_y = \frac{\partial^2 U}{\partial T \, \partial Y} + \left(\frac{\partial \zeta}{\partial T}\right)_Y.$$

It is clear from this, in view of (1.41), that when it is applied to Eq. (1.46), the condition (1.43) is not satisfied or else ($\partial N/\partial x$)$_y$ differs from ($\partial M/\partial y$)$_x$ by an amount ($\partial \zeta/\partial T$)$_Y$. Consequently the differential dQ is not a total differential.

A similar conclusion can be reached for another process function, the work

$$dL = \zeta \, dY. \qquad (1.19)$$

The work L is a function of two variables: the generalized force ζ and the generalized coordinate Y. We can evidently write

$$dL = M d\zeta + N dY \qquad (1.47)$$

for the differential of this function $L = f(\zeta, Y)$, with the understanding that here $x = \zeta$; and $y = Y$.

We can check whether Eq. (1.43) is satisfied for the function $L = f(\zeta, Y)$. Comparing (1.19) and (1.47) we see that $M = 0$ and $N = \zeta$ and consequently

$$\left(\frac{\partial M}{\partial y}\right)_x = 0; \quad \left(\frac{\partial N}{\partial x}\right)_y = 1.$$

From this we see that

$$\left(\frac{\partial M}{\partial y}\right)_x \neq \left(\frac{\partial N}{\partial x}\right)_y,$$

and thus the differential dL is also not a total differential.

When $z = $ const it follows from Eq. (1.40) that

$$\left(\frac{\partial z}{\partial y}\right)_x \left(\frac{\partial y}{\partial x}\right)_z \left(\frac{\partial x}{\partial z}\right)_y = -1. \tag{1.48}$$

Naturally if a quantity z is a function of two variables x and y, i.e., $z = f(x, y)$, on this basis we can regard x as a function of the variables y and z, i.e., $x = \varphi(y, z)$, while y can be regarded as a function of x and z, i.e., $y = \psi(x, z)$. Equation (1.48) uniquely relates the derivatives of these three functions. It allows us to relate any three thermodynamic state functions. Thus we find for p, v, and T from (1.46)

$$\left(\frac{\partial p}{\partial T}\right)_v \left(\frac{\partial T}{\partial v}\right)_p \left(\frac{\partial v}{\partial p}\right)_T = -1,$$

and for p, T, and s

$$\left(\frac{\partial p}{\partial T}\right)_s \left(\frac{\partial T}{\partial s}\right)_p \left(\frac{\partial s}{\partial p}\right)_T = -1,$$

while for i, u, and T

$$\left(\frac{\partial i}{\partial u}\right)_T \left(\frac{\partial u}{\partial T}\right)_i \left(\frac{\partial T}{\partial i}\right)_u = -1,$$

etc.

We can find yet another important relation from Eq. (1.40):

$$\left(\frac{\partial z}{\partial x}\right)_\eta = \left(\frac{\partial z}{\partial x}\right)_y + \left(\frac{\partial z}{\partial y}\right)_x \left(\frac{\partial y}{\partial x}\right)_\eta. \tag{1.49}$$

This equation allows us to establish a relation between partial

derivatives of the same quantity calculated with different parameters held constant.

If the system is composed of a pure material† and undergoes
a single form of work (for example, work of expansion) then the
state of the material in such a system will be uniquely determined
if two parameters are given. Any other parameter is a unique
function of the two given parameters. Consequently for such a system any three state parameters (p, v, and T, for example) of a
pure material are uniquely related to each other.

The equation which relates these parameters to each other
is called the equation of state of the given material. For the given
system the equation of state contains three variables.

If a system undergoes two forms of work it is clear that the
state of pure material in such a system will be determined not by
two but by three parameters (there are correspondingly four parameters in the equation of state), etc. The state of the material
in a system undergoing n forms of work is determined by (1 + n)
parameters and the equation of state for such a system contains
(2 + n) variables.

In what follows we will as a rule be interested in systems
which undergo no more than two forms of work (one of which is
the work of expansion done by the system against external pressure).

---

†In this book we will only consider systems composed of pure materials.

Equilibrium Thermodynamic Systems Which Undergo Other Forms of Work in Addition to Work of Expansion

2.1. Criteria for Equilibrium in Thermodynamic Systems

The second law of thermodynamics establishes important criteria for equilibrium in isolated thermodynamic systems. By isolated we mean a system which does not exchange heat or work with the external medium. Consequently in such a system the internal energy U, the volume V, and the generalized coordinate X (which corresponds to a form of work other than work of expansion) are constant.

In agreement with the second law of thermodynamics, the entropy of an isolated system goes to a maximum. The entropy of an isolated system has its greatest possible value in equilibrium; in other words dS = 0 in an isolated system in equilibrium. Indeed for an isolated system dU = 0, dV = 0, and dX = 0, and from Eq. (1.28)

$$T dS \geqslant dU + p dV + \xi dX$$

we find

$$dS \geqslant 0. \tag{2.1}$$

This condition determines the evolution of an isolated system. Here the inequality corresponds to a nonequilibrium state of the system where the system is still "on the way" to the equilibrium state (all nonequilibrium processes are irreversible) and the equal sign corresponds to a system already in equilibrium.

Thus in an equilibrium state of an isolated system

$$dS = 0. \qquad (2.2)$$

If the system is not isolated from the external medium and can interact with the surrounding medium in some way (i.e., if it is coupled to the medium), the equilibrium conditions will be different from (2.2). The equilibrium condition will depend on the conditions for the interaction of the system with the surrounding medium.

For a system whose only form of work is work of expansion (for brevity we will refer to such a system as "simple") we are most interested in four types of interaction conditions (or, as we say, coupling conditions) between the system and the surrounding medium:

(1) The volume of the system is fixed but the system can exchange heat with the surrounding medium in such a way that the entropy of the system remains constant:

$$V = \text{const}, \ S = \text{const}.$$

(2) The system can exchange both heat and work of expansion with the surrounding medium in such a way that the pressure and entropy of the system remain constant:

$$p = \text{const}, \ S = \text{const}.$$

(3) The volume of the system remains constant but it can exchange heat with the surrounding medium in such a way that the temperature of the system remains constant (isochoric–isothermal system):

$$V = \text{const}, \ T = \text{const}.$$

(4) The system can exchange both heat and work with the surrounding medium under conditions where the pressure and temperature of the system remain fixed (isobaric–isothermal system):

$$p = \text{const}, \ T = \text{const}.$$

The criteria for equilibrium for each of these four cases of interaction between the system and the medium are as follows.

(1) Interaction conditions V = const and S = const. It is evident from Eq. (1.29), rewritten in the form

$$dU \leqslant TdS - pdV, \qquad (2.3)$$

that the evolution of the system under consideration, in which dV = 0 and dS = 0, is given by the condition

$$dU \leqslant 0. \qquad (2.4)$$

In other words, the internal energy of the system decreases in approaching the equilibrium state, reaching a minimum in the equilibrium state. Thus in the equilibrium state

$$dU = 0. \qquad (2.5)$$

(2) Interaction conditions p = const and S = const. Applying the Legendre transformation

$$pdV = d(pV) - Vdp \qquad (2.6)$$

to the quantity pdV and using

$$I = U + pV, \qquad (2.7)$$

we can transform Eq. (1.29) to the following form:

$$dI \leqslant TdS + Vdp. \qquad (2.8)$$

It is thus evident that any process in this system in which dp = 0 and dS = 0 occurs in such a way that the condition

$$dI \leqslant 0 \qquad (2.9)$$

is satisfied, i.e., the enthalpy of the system decreases in approaching the equilibrium state, reaching a minimum in the equilibrium state. Thus in the equilibrium state

$$dI = 0. \tag{2.10}$$

(3) **Interaction conditions V = const and T = const.** Applying the Legendre transformation

$$TdS = d(TS) - SdT \tag{2.11}$$

to TdS and setting

$$F = U - TS, \tag{2.12}$$

we find the well-known thermodynamic function called the free energy (isochoric–isothermal potential). We then transform Eq. (1.29) to the following form

$$dF \leqslant -SdT - pdV. \tag{2.13}$$

It is evident from this that the evolution of a system in which dT = 0 and dV = 0 is given by the condition

$$dF \leqslant 0, \tag{2.14}$$

i.e., the free energy of the system decreases in approaching the equilibrium state, reaching a minimum in this state. Thus in the equilibrium state

$$dF = 0. \tag{2.15}$$

(4) **Interaction conditions p = const and T = const.** Replacing pdV in Eq. (1.29) by pdV in relation (2.6) and TdS by (2.11) and setting

$$\Phi = I - TS \tag{2.16}$$

(note that this thermodynamic function is called the isobaric–isothermal potential), we find

$$d\Phi \leqslant -SdT + Vdp. \tag{2.17}$$

The evolution of this system, in which dT = 0 and dp = 0, is determined by

$$d\Phi \leqslant 0, \tag{2.18}$$

i.e., the isobaric–isothermal potential of the system decreases in approaching the equilibrium state, reaching a minimum in this state. Thus in the equilibrium state

$$d\Phi = 0. \tag{2.19}$$

For an isolated "simple" system (U = const and V = const), of course, the previously obtained criterion (2.2) is valid.

These are the equilibrium criteria for thermodynamic systems in which the only form of work is work of expansion.

When a system undergoes other forms of work in addition to work of expansion the equilibrium criteria will be somewhat different. By methods similar to those used above, from Eq. (1.27)

$$TdS \geqslant dU + pdV + dL^*,$$

where

$$dL^* = \xi dX,$$

we find the following results when the system interacts with the surrounding medium.

(1) For interaction conditions V = const and S = const,

$$dU + dL^* \leqslant 0, \tag{2.20}$$

i.e., in the equilibrium state

$$dU = -dL^*. \tag{2.21}$$

(2) For interaction conditions p = const and S = const,

$$dI + dL^* \leqslant 0, \tag{2.22}$$

i.e., in the equilibrium state

$$dI = -dL^*. \tag{2.23}$$

(3) For interaction conditions V = const and
T = const,

$$dF + dL^* \leqslant 0, \tag{2.24}$$

i.e., in the equilibrium state

$$dF = -dL^*. \tag{2.25}$$

(4) For interaction conditions p = const and
T = const,

$$d\Phi + dL^* \leqslant 0, \tag{2.26}$$

i.e., in the equilibrium state

$$d\Phi = -dL^*. \tag{2.27}$$

As for systems which interact with the surrounding medium under conditions U = const and V = const, it should be clear to the reader than when the system undergoes other forms of work besides work of expansion these conditions still do not guarantee the isolated nature of the system (for this it would also be necessary to have X = const) so that, as we see from Eq. (1.27), in this case

$$TdS \geqslant dL^* \tag{2.28}$$

or, what is the same,

$$dS \geqslant \frac{\xi}{T} dX, \tag{2.29}$$

so that in the equilibrium state

$$dS = \frac{\xi}{T} dX. \tag{2.30}$$

We can see from Eqs. (2.21), (2.23), (2.25), and (2.27) that the work which can be performed by a system in equilibrium under given conditions of coupling with the medium (after subtracting the work of expansion) equals the decrease in the corresponding characteristic function.[†]

Because of this, by a well-known analogy with mechanics, characteristic functions are referred to as thermodynamic potentials.

We sometimes call Φ the free enthalpy by analogy with the free energy F. The origin of the terms "free energy" and "free enthalpy" is as follows. It is evident from (2.12) that the expression for the internal energy of the system can be put in the following form:

$$U = F + TS.$$

As shown above, in an isochoric–isothermal system in equilibrium work L^* can only be accomplished due to a drop in F [Eq. (2.22)]. Consequently in such a system not all of the work can be transformed into internal energy but only the "free" part, F. The quantity TS, which is sometimes called the bound energy, cannot be transformed into work. Analogously, in an isobaric–isothermal system in equilibrium, as we see from Eq. (2.27), work can only be accomplished due to a decrease in Φ, the "free" part of the system enthalpy

$$I = \Phi + TS.$$

We now consider the criteria for equilibrium in a system for different conditions of interaction with the surrounding medium including, in addition to the conditions enumerated above, constant values of the ξ or X which characterize the given form of work.

For an isolated system (U = const, V = const, X = const) the condition for equilibrium, as noted at the beginning of this section, is a maximum in the entropy of the system [see Eqs. (2.1) and (2.2)].

Among other conditions for the interaction between the system and the surrounding medium it is of interest to consider systems in which

(1) V = const, S = const, and X = const,
(2) p = const, S = const, and ξ = const,
(3) V = const, T = const, and X = const,
(4) p = const, T = const, and ξ = const.

[†]The quantities U, I, F, and Φ are called characteristic functions. Characteristic functions have the following distinct properties: If the characteristic function is known in terms of the corresponding variables (different for each characteristic function), it can be used to calculate any thermodynamic quantity. Characteristic functions are additive.

We will use the methods given above to investigate the criteria for equilibrium in such systems.

(1) Interaction conditions V = const, S = const, and X = const. From Eqs. (1.28), rewritten in the form

$$dU \leqslant TdS - pdV - \xi dX,$$

we find that the evolution of a system in which $dV = 0$, $dS = 0$, and $dX = 0$ is given by the condition

$$dU \leqslant 0, \tag{2.31}$$

i.e., in approaching the equilibrium state the internal energy of the system decreases, reaching a minimum in the equilibrium state. Thus, in the equilibrium state

$$dU = 0. \tag{2.32}$$

Thus the criteria for equilibrium in a system with V = const, S = const, and X = const are the same as for the "simple" system with the condition that V = const and S = const [Eqs. (2.4) and (2.5)].

(2) Interaction conditions p = const, S = const, and ξ = const. Since

$$I = U + pV,$$

Eq. (1.28) can be written in the form

$$dI \leqslant TdS + Vdp - \xi dX. \tag{2.33}$$

The quantity ξdX which enters this equation can be represented as follows using the Legendre transformation:

$$\xi dX = d(\xi X) - X d\xi. \tag{2.34}$$

Thus we find from (2.33)

$$d(I + \xi X) \leqslant TdS + Vdp + X d\xi. \tag{2.35}$$

We see from this that in a system in which $dp = 0$, $dS = 0$, and $d\xi = 0$ the direction of a process is given by the condition

$$d(I + \xi X) \leqslant 0,$$ (2.36)

i.e., in the equilibrium state

$$d(I + \xi X) = 0.$$ (2.37)

As we see, our criteria for equilibrium in the system with p = const, S = const, and ξ = const differ from the equilibrium criteria for a "simple" system in which only p = const and S = const [Eq. (2.9)].

What physical significance is to be attached to the quantity (I + ξX) which we have introduced in the derivation? We recall the enthalpy I defined by Eq. (2.7) represents the energy of an "extended" system (the internal energy U plus the potential energy pV which is given to the system by a piston with a force on it). We can see by analogy with this that for the systems we have considered which can undergo work against some force ξ in addition to work of expansion against external pressure, the energy of the extended system can be written in the form U + pV + ξX. This quantity can be regarded as the enthalpy of the system under study; we will denote the enthalpy of a system in which dL = pdV + ξdX by I*:

$$I^* = U + pV + \xi X.$$ (2.38)

It can be seen from this that the enthalpy I* is related to the "usual" enthalpy I as follows:

$$I^* = I + \xi X.$$ (2.39)

Relation (2.37) can be written in the form

$$dI^* = 0.$$ (2.40)

This relation is similar to the equilibrium condition for a "simple" system in which p = const and S = const (dI = 0), but with the difference that the quantity I* which enters (2.40) differs from the "ordinary" enthalpy I.

(3) Interaction conditions V = const, T = const, and X = const. In view of the fact that

$$F = U - TS,$$

we write Eq. (1.28) in the form

$$dF \leqslant -SdT - pdV - \xi dX.$$ (2.41)

It can be seen from this that the evolution of the system is governed by the condition

$$dF \leqslant 0, \qquad\qquad (2.42)$$

i.e., in the equilibrium state

$$dF = 0. \qquad\qquad (2.43)$$

This condition agrees with the equilibrium condition for a "simple" system in which V = const, T = const, and ξ = const [Eq. (2.14)].

(4) Interaction conditions p = const, T = const, and ξ = const. In view of the fact that

$$\Phi = I - TS,$$

Eq. (1.28) can be represented in the form

$$d\Phi \leqslant -SdT + Vdp - \xi dX. \qquad\qquad (2.44)$$

Using the transformation (2.34), we find from (2.44)

$$d(\Phi + \xi X) \leqslant -SdT + Vdp + Xd\xi.$$

It is evident from this that in a system in which dV = 0, dT = 0, and dX = 0, all processes occur in such a way that

$$d(\Phi + \xi X) \leqslant 0, \qquad\qquad (2.45)$$

i.e., in the equilibrium state

$$d(\Phi + \xi X) = 0. \qquad\qquad (2.46)$$

This condition differs from the equilibrium condition for a "simple" system with p = const and T = const (dΦ = 0).

Taking (2.16) into account, (Φ + ξX) can be written as

$$\Phi + \xi X = U + pV - TS + \xi X = I - TS + \xi X, \qquad\qquad (2.47)$$

and since, in agreement with (2.39),

$$I^* = I + \xi X,$$

it follows that

$$\Phi + \xi X = I^* - TS. \qquad (2.48)$$

By analogy with (2.39) we will use the notation

$$\Phi^* = I^* - TS. \qquad (2.49)$$

The quantity Φ^* can be regarded as the isobaric—isothermal potential for this system. Φ^* is related to the "ordinary" isobaric—isothermal potential Φ by the obvious equation

$$\Phi^* = \Phi + \xi X. \qquad (2.50)$$

Now relation (2.46) can be written in the form

$$d\Phi^* = 0. \qquad (2.51)$$

In such a form this condition is similar to the equilibrium condition for a "simple" system with p = const and T = const (dΦ = 0) but with the difference that instead of the "ordinary" isobaric—isothermal potential Φ, Eq. (2.51) contains Φ^*.

Such are the criteria for equilibrium in thermodynamic systems which undergo other kinds of work besides work of expansion, for interaction conditions with the surrounding medium such that,

TABLE 2.1

Interaction conditions between the system and the surrounding medium	Form of work undergone by system	
	$dL = pdV$	$dL = pdV + \xi dX$
$U = \text{const}, \ V = \text{const}$	$dS \geqslant 0$	$dS \geqslant \dfrac{\xi}{T} dX$
$U = \text{const}, \ V = \text{const}, \ X = \text{const}$	—	$dS \geqslant 0$
$S = \text{const}, \ V = \text{const}$	$dU \leqslant 0$	$dU \leqslant -dL^*$
$S = \text{const}, \ V = \text{const}, \ X = \text{const}$	—	$dU \leqslant 0$
$S = \text{const}, \ p = \text{const}$	$dI \leqslant 0$	$dI \leqslant -dL^*$
$S = \text{const}, \ p = \text{const}, \ \xi = \text{const}$	—	$d(I + \xi X) \leqslant 0$
$T = \text{const}, \ V = \text{const}$	$dF \leqslant 0$	$dF \leqslant -dL^*$
$T = \text{const}, \ V = \text{const}, \ X = \text{const}$	—	$dF \leqslant 0$
$T = \text{const}, \ p = \text{const}$	$d\Phi \leqslant 0$	$d\Phi \leqslant -dL^*$
$T = \text{const}, \ p = \text{const}, \ \xi = \text{const}$	—	$d(\Phi + \xi X) \leqslant 0$

in addition to the usual interaction conditions, either ξ or X is held constant.

The conditions of evolution obtained above for various conditions on the interaction between a thermodynamic system and the surrounding medium and for various kinds of work performed by the system are given in Table 2.1.

2.2. The Chemical Potential

The concept of the chemical potential plays an important role in the analysis of various processes which take place in thermodynamic systems (in particular in the analysis of phase transformations).

The chemical potential of material in a "simple" system (see Section 2.1) is the weight-specific isobaric—isothermal potential

$$\varphi = i - Ts. \qquad (2.52)$$

Here i and s are the weight-specific enthalpy and entropy.

In view of (2.39) it is evident that for systems which undergo other forms of work in addition to work of expansion, we should define the chemical potential as

$$\varphi^* = i + \xi x - Ts, \qquad (2.53)$$

where x is the weight-specific generalized X coordinate, or, what is the same,

$$\varphi^* = i^* - Ts. \qquad (2.54)$$

The chemical potential occupies a very special position among the other weight-specific thermodynamic potentials — the internal energy, enthalpy, and free energy. The reasons for this are as follows.

Above in considering the criteria for equilibrium in thermodynamic systems under various interaction conditions with the medium we tacitly assumed that the amount of material G in the system remained unchanged. However, in solving many problems (in particular for the analysis of phase equilibrium conditions) it is useful to establish how the thermodynamic potential of the system changes when a certain amount of material dG is taken from (or added to) the system (it is to be understood that we speak of adding material which has the same state parameters as the material already in the system). In other words, we must find the quantities

$\left(\dfrac{\partial U}{\partial G}\right)_{S, V, X}$ for a system with $S=$const, $V=$const, $X=$const;

$\left(\dfrac{\partial I}{\partial G}\right)_{S, p, \xi}$ for a system with $S=$const, $p=$const, $\xi=$const;

$\left(\dfrac{\partial F}{\partial G}\right)_{T, V, X}$ for a system with $T=$const, $V=$const, $X=$const;

$\left(\dfrac{\partial \Phi}{\partial G}\right)_{T, p, \xi}$ for a system with $T=$const, $p=$const, $\xi=$const.

We consider a system with S = const, V = const, and X = const. Evidently the internal energy U of the system can be written as

$$U = Gu, \qquad (2.55)$$

where G is the weight of material in the system and u is the weight-specific internal energy.

It follows from this that

$$dU = Gdu + udG. \qquad (2.56)$$

Since, as we see from (1.28b),

$$du = Tds - pdv - \xi dx,$$

Eq. (2.56) can be transformed to the form

$$dU = TGds - pGdv - \xi Gdx + udG. \qquad (2.57)$$

Using the Legendre transformations

$$Gds = d(Gs) - sdG, \qquad (2.58)$$

$$Gdv = d(Gv) - vdG, \qquad (2.59)$$

$$Gdx = d(Gx) - xdG \qquad (2.60)$$

and the definitions

$$Gs = S, \quad Gv = V \text{ and } Gx = X,$$ (2.61)

we find from (2.57) that

$$dU = (u + pv + \xi x - Ts) dG + TdS - pdV - \xi dX.$$ (2.62)

Since S = const, V = const, and X = const in the system, using

$$i = u + pv$$ (2.7a)

and (2.53) we find

$$dU = \varphi^* dG;$$ (2.63)

consequently

$$\left(\frac{\partial U}{\partial G} \right)_{S, V, X} = \varphi^*.$$ (2.63a)

We now assume that the system obeys S = const, p = const, and ξ = const. From the obvious relation for the enthalpy I^* of the system

$$I^* = Gi^*,$$ (2.64)

where i^* is the weight-specific enthalpy of the system, it follows that

$$dI^* = Gdi^* + i^* dG.$$ (2.65)

As regards i^* it is obvious from (2.38) and (2.39) that

$$i^* = u + pv + \xi x = i + \xi x.$$ (2.66)

Using (2.66) and

$$pdv = d(pv) - vdp$$ (2.67)

and

$$\xi dx = d(\xi x) - xd\xi,$$ (2.68)

we find from (1.28b)

$$di^* = Tds + vdp + xd\xi. \tag{2.69}$$

Using this relation, Eq. (2.65) can be transformed to the form

$$dI^* = TGds + vGdp + xGd\xi + i^*dG. \tag{2.70}$$

Using (2.58) and (2.61) we find

$$dI^* = (i^* - Ts)dG + TdS + Vdp + Xd\xi. \tag{2.71}$$

Since S = const, p = const, and ξ = const in the system of interest, using (2.54) we find for the given system

$$dI^* = \varphi^* dG; \tag{2.72}$$

consequently,

$$\left(\frac{\partial I^*}{\partial G}\right)_{S,\,p,\,\xi} = \varphi^*. \tag{2.72a}$$

We now consider a system with T = const, V = const, and X = const. From the relation for the free energy of the system

$$F = Gf, \tag{2.73}$$

where f is the weight-specific free energy

$$f = u - Ts, \tag{2.74}$$

we find

$$dF = Gdf + fdG. \tag{2.75}$$

It follows from (2.74) that

$$df = du - Tds - sdT \tag{2.76}$$

or, using (1.28b),

$$df = -pdv - \xi dx - sdT. \tag{2.77}$$

Substituting this relation into (2.75), we find that

$$dF = -pGdv - \xi Gdx - sGdT + f dG. \qquad (2.78)$$

Using the transformations (2.59) and (2.60) and Eq. (2.61), we find

$$dF = (f + pv + \xi x)\, dG - pdV - \xi dX - SdT. \qquad (2.79)$$

Furthermore, as we see from (2.53), (2.54), (2.7a), and (2.74),

$$\varphi^* = i^* - Ts = i + \xi x - Ts = u + pv + \xi x - Ts = f + pv + \xi x. \qquad (2.80)$$

Since we have T = const, V = const, and X = const in the system, in view of (2.80) we find from (2.79)

$$dF = \varphi^* dG; \qquad (2.81)$$

consequently

$$\left(\frac{\partial F}{\partial G} \right)_{T,\,V,\,X} = \varphi^*. \qquad (2.81a)$$

Finally we consider a system with T = const, p = const, and ξ = const. From the relation for the potential Φ^* of the system

$$\Phi^* = G\varphi^* \qquad (2.82)$$

it follows that

$$d\Phi^* = Gd\varphi^* + \varphi^* dG. \qquad (2.83)$$

From Eq. (2.80)

$$\varphi^* = u + pv + \xi x - Ts$$

we find

$$d\varphi^* = du + pdv + vdp + \xi dx + xd\xi - Tds - sdT. \qquad (2.84)$$

Using Eq. (1.28b)

$$Tds = du + pdv + \xi dx$$

we find

$$d\varphi^* = vdp + xd\xi - sdT. \tag{2.85}$$

Substituting this relation into (2.83) and taking account of (2.61), we find

$$d\Phi^* = Vdp + Xd\xi - SdT + \varphi^*dG. \tag{2.86}$$

Since T = const, p = const, and ξ = const in the system we find from (2.86)

$$d\Phi^* = \varphi^*dG; \tag{2.87}$$

consequently

$$\left(\frac{\partial\Phi^*}{\partial G}\right)_{T,\,p,\,\xi} = \varphi^*. \tag{2.87a}$$

Thus, as follows from Eqs. (2.63a), (2.72a), (2.81a), and (2.87a),

$$\left(\frac{\partial U}{\partial G}\right)_{S,V,X} = \left(\frac{\partial I^*}{\partial G}\right)_{S,p,\xi} = \left(\frac{\partial F}{\partial G}\right)_{T,V,X} = \left(\frac{\partial\Phi^*}{\partial G}\right)_{T,p,\xi} = \varphi^*. \tag{2.88}$$

In other words, for each of the systems under consideration the derivative of the corresponding characteristic function with respect to the amount of material in the system equals Φ^*. Consequently the weight-specific potential Φ^* occupies a special position among other weight-specific characteristic functions; this quantity has a remarkable property – it allows one to calculate the change in the characteristic functions for any system with a change in the amount of material in the system; for this reason Φ^* is called the chemical potential.

We should note one further circumstance. As noted above, for an isolated system (U = const, V = const, X = const) the quantity which characterizes the equilibrium state is the entropy (in equilibrium the entropy is a maximum). The entropy is not a characteristic function. However, it is of interest to note that the derivative of the entropy of the system with respect to G is also related to Φ^*. Indeed, from the obvious relation

$$S = Gs \tag{2.89}$$

for the entropy of the system, it follows that

$$dS = Gds + sdG. \tag{2.90}$$

Replacing ds in this relation using (1.28b), we find

$$dS = \frac{1}{T}(G du + pG dv + \xi G dx) + s dG.$$ (2.91)

Using transformations (2.59) and (2.60) and

$$G du = d(Gu) - u dG,$$ (2.92)

and using (2.61) and

$$Gu = U$$ (2.93)

we find from (2.91)

$$dS = \frac{1}{T}(dU + p dV + \xi dX) - \frac{1}{T}(u + pv + \xi x - Ts) dG$$ (2.94)

or, in view of (2.80),

$$dS = \frac{1}{T}(dU + p dV + \xi dX) - \frac{\varphi^*}{T} dG.$$ (2.95)

Since dU = 0, dV = 0, and dX = 0 in an isolated system, for the system of interest

$$dS = -\frac{\varphi^*}{T} dG;$$ (2.96)

consequently

$$\left(\frac{\partial S}{\partial G}\right)_{U, V, X} = -\frac{\varphi^*}{T}.$$ (2.96a)

Equation (2.95) can be written in the following form:

$$T dS = dU + p dV + \xi dX - \varphi^* dG.$$ (2.95a)

Evidently this relation is the combined equation for the first and second laws of thermodynamics for systems with a variable amount of material.

In conclusion we formulate certain relations which we will use in what follows. From Eq. (2.85)

$$d\varphi^* = v dp + x d\xi - s dT$$

it follows that

$$\left(\frac{\partial \varphi^*}{\partial T}\right)_{p, \xi} = -s,$$ (2.97)

$$\left(\frac{\partial \varphi^*}{\partial p}\right)_{T, \xi} = v,$$ (2.98)

$$\left(\frac{\partial \varphi^*}{\partial \xi}\right)_{T, p} = x.$$ (2.99)

2.3. Equilibrium Conditions in an Isolated Homogeneous System

Of the various thermodynamic systems which differ from each other in having different interaction conditions with the surrounding medium, the treatment of the equilibrium conditions in an isolated thermodynamic system is of greatest practical interest.

We consider a system such as schematically illustrated in Fig. 2.1. We conceptually divide this system into two parts (or, as we will say, subsystems) 1 and 2 and seek the conditions for equilibrium between these two subsystems.

Since the system as a whole is isolated,

$$U_{\text{syst}} = \text{const}, \; V_{\text{syst}} = \text{const} \text{ and } X_{\text{syst}} = \text{const}. \qquad (2.100)$$

In addition it is clear that we can think of an infinitesimally small process within the isolated system in which each of the subsystems changes only one of these quantities, any pair, or all of them simultaneously. We let the volume of the first subsystem vary by an amount dV_1, the X coordinate by an amount dX_1, and the internal energy by an amount dU_1, while for the second subsystem these quantities are, respectively, dV_2, dX_2, and dU_2.

It is further evident that

$$V_{\text{syst}} = V_1 + V_2 \qquad (2.101)$$

and

$$U_{\text{syst}} = U_1 + U_2; \qquad (2.102)$$

and we will assume that

$$X_{\text{syst}} = X_1 + X_2 \qquad (2.103)$$

by analogy with (2.101).[†]

Using (2.100) it follows that

$$dV_{\text{syst}} = dV_1 + dV_2 = 0, \qquad (2.104)$$

[†]With regard to the range of applicability of Eq. (2.103), see p. 38.

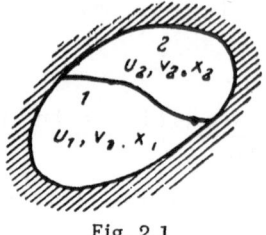

Fig. 2.1

$$dX_{\text{syst}} = dX_1 + dX_2 = 0 \qquad (2.105)$$

and

$$dU_{\text{syst}} = dU_1 + dU_2 = 0, \qquad (2.106)$$

i.e.,

$$dV_1 = -dV_2. \qquad (2.107)$$

$$dX_1 = -dX_2 \qquad (2.108)$$

and

$$dU_1 = -dU_2. \qquad (2.109)$$

In other words, there is some decrease in the volume (or X, or internal energy) of the first subsystem and a matching increase in the volume (or X, or internal energy) of the second subsystem.

We previously established an important criterion for equilibrium in an isolated system: it was shown that in a state of thermodynamic equilibrium the entropy of an isolated system preserves a constant (maximum) value, i.e., in the equilibrium state

$$dS_{\text{syst}} = 0. \qquad (2.110)$$

Since the entropy is an additive quantity, it is evident that in our case we have

$$S_{\text{syst}} = S_1 + S_2, \qquad (2.111)$$

where S_1 and S_2 are the entropies of the first and second subsystems respectively. In agreement with (2.110) we find from this that

$$dS_{\text{syst}} = dS_1 + dS_2 = 0. \qquad (2.112)$$

Furthermore from Eq. (1.28a)

$$TdS = dU + pdV + \xi dX$$

we find

$$dS = \frac{1}{T} dU + \frac{p}{T} dV + \frac{\xi}{T} dX$$

(2.113)

Consequently for the first subsystem

$$dS_1 = \frac{1}{T_1} dU_1 + \frac{p_1}{T_1} dV_1 + \frac{\xi_1}{T_1} dX_1,$$

(2.114)

while for the second

$$dS_2 = \frac{1}{T_2} dU_2 + \frac{p_2}{T_2} dV_2 + \frac{\xi_2}{T_2} dX_2.$$

(2.115)

Substituting dS_1 and dS_2 from relations (2.114) and (2.115) into Eq. (2.112) and using (2.107) and (2.109), we find

$$\left(\frac{1}{T_1} - \frac{1}{T_2}\right) dU_1 + \left(\frac{p_1}{T_1} - \frac{p_2}{T_2}\right) dV_1 + \left(\frac{\xi_1}{T_1} - \frac{\xi_2}{T_2}\right) dX_1 = 0.$$

(2.116)

Above we noted that the internal energy U, volume V, and generalized coordinate X can be varied independently of each other, i.e., they can represent, for example, a process in which the volume of each of the subsystems varies while their internal energies and X values remain fixed; similarly we can have processes in which a change in U does not imply changes in V and X and a change in X does not imply changes in U and V. In other words, the differentials dU_1, dV_1, and dX_1 are independent in principle. If this is so, then it is evident that for the left-hand side of (2.116) to be identically equal to zero it is necessary that the coefficients of the differentials dU_1, dV_1, and dX_1 in this equation be zero independently of each other, i.e., that

$$\frac{1}{T_1} - \frac{1}{T_2} = 0,$$

(2.117)

$$\frac{p_1}{T_1} - \frac{p_2}{T_2} = 0,$$

(2.118)

and

$$\frac{\xi_1}{T_1} - \frac{\xi_2}{T_2} = 0.$$

(2.119)

It follows from (2.117) that

$$T_1 = T_2,$$

(2.120)

while from (2.118) and (2.119) using (2.120) we find that

$$p_1 = p_2$$

(2.121)

and
$$\xi_1 = \xi_2. \tag{2.122}$$

Evidently we arrive at the same result independent of how we conceptually divide the system into two subsystems. Thus we conclude that in an isolated system which undergoes work against a certain force ξ in addition to work against pressure, in equilibrium the temperature, pressure, and force ξ are identical in all parts of the system.

It is clear from this derivation that for "ordinary" isolated systems (which only undergo work of expansion) equilibrium will be characterized by (2.120) and (2.121).

Naturally a question arises: Are our conclusions of a general character for any isolated system or have we used some assumptions in carrying out the derivation that limit the scope of these conclusions?

Actually we have tacitly introduced a number of limitations in the derivation. In the first place the conclusions are only valid for generalized coordinates which are extensive and for which the additivity condition (2.103) is valid, as for the volume V. It should be noted that this condition is not satisfied for a number of generalized coordinates. Thus, for example, for a system located in a gravitational field the condition (2.103) makes no sense. In this case the generalized coordinate is the height h, and it should be understood that the height of the center of mass of the system is not the sum of the heights of the centers of mass of the subsystems. Thus Eq. (2.121) is not satisfied for a gas or liquid located in a gravitational field; the pressure in the gas (liquid) column increases with decreasing height (see Chapter 7).

In what follows in treating various systems we will each time mention whether the additivity condition is satisfied for the generalized coordinate characterizing the properties of this system.

In the second place we have assumed in the derivation that the dividing surface between the subsystems does not have any features which would have to be considered. This assumption is valid when we consider single-phase systems. When the material in the subsystems is in different phases, strictly speaking it is necessary to take account of surface forces, which, as we will see below (Chapter 6), have a number of special properties. In this case we would have to add yet another term to the right-hand side of Eq. (2.102) – the energy of the surface layer. It should be noted that when the separation surface is planar, all three equilibrium conditions for the system, (2.120), (2.121), and (2.122), remain unchanged. For curvilinear bounding surfaces the pressures in phases in mutual equilibrium will be different (this will be shown below in Chapter 6).

Finally, it should be stressed that Eqs. (2.120)-(2.122) were obtained for isolated systems. In fact, for a number of cases which we will consider in this book the idea of an isolated system makes no practical sense. Indeed since (for example) at the present time we know no way of shielding things from gravitation it makes no sense to speak of isolating ordinary systems from the action of a gravitational field, and although a system can be isolated in a magnetic sense by surrounding it by a superconducting shield which is impermeable to a magnetic field, for most practical cases such isolation conditions are not applicable.

2.4. The Conditions for Phase Equilibrium

In Section 2.3 we considered the conditions for phase equilibrium in an isolated homogeneous system and found that in such a system (within certain limitations) the temperature, pressure, and force ξ are the same in all parts of the system.

We now consider isolated systems composed of two (or more) phases† and find the general conditions for phase equilibrium.

We consider an isolated thermodynamic system composed of two subsystems 1 and 2. We use the same methods as in Section 2.3. The only difference is that now the material in the subsystems is in two different phases; the first subsystem consists of one phase, the second subsystem consists of the other phase, and the amount of material in the subsystems can vary (transformation of material from one phase into the other).

Thus we consider a system for which

$$V_{syst} = \text{const}, \ X_{syst} = \text{const}, \ U_{syst} = \text{const}, \ \text{and} \ G_{syst} = \text{const}$$

(here G_{syst} is the total amount of material in the system).

---

†By "phases" we mean homogeneous regions in a heterogeneous system.

As shown in Section 2.3, evidently

$$V_{\text{syst}} = V_1 + V_2, \tag{2.123}$$

$$G_{\text{syst}} = G_1 + G_2, \tag{2.124}$$

$$U_{\text{syst}} = U_1 + U_2, \tag{2.125}$$

assuming also that

$$X_{\text{syst}} = X_1 + X_2 \tag{2.126}$$

(here subscripts 1 and 2 refer, respectively, to the first and second phases).

Equations (2.123), (2.125), and (2.126) are similar, respectively, to relations (2.101), (2.102), and (2.103), but with the difference that here the different subsystems correspond to different phases and transfer of material from one subsystem into the other occurs via a phase transformation.

We note that in the present case, as previously, we do not take account of the energy of the surface layer at the interface between phases, which in general has a certain amount of additional (surface) energy. However, as noted above, for planar interfaces this does not affect the validity of the conditions for phase equilibrium which will be obtained below.

Since V_{syst} = const, X_{syst} = const, G_{syst} = const, and U_{syst} = const it follows from (2.123)–(2.126) that

$$dV_1 = -dV_2, \tag{2.127}$$

$$dG_1 = -dG_2, \tag{2.128}$$

$$dU_1 = -dU_2, \tag{2.129}$$

$$dX_1 = -dX_2. \tag{2.130}$$

As we know, in equilibrium the entropy of an isolated system remains constant (at a maximum) so that

$$dS_{\text{syst}} = 0. \tag{2.110}$$

Furthermore it is clear that since the entropy is an additive quantity,

$$S_{syst} = S_1 + S_2, \tag{2.131}$$

whence, using Eq. (2.110), we find

$$dS_{syst} = dS_1 + dS_2. \tag{2.132}$$

We now consider the expression for the differential entropy for each of the subsystems. It is clear that in contrast to the assumption of Section 2.3 for the case where the amount of material in the subsystem was constant, we now should regard the entropy of the subsystem as a function not only of U_1, V_1, and X_1 but also of the amount of material in the subsystem, G_1, i.e.,

$$S_1 = f(U_1, V_1, X_1, G_1). \tag{2.133}$$

In agreement with Eq. (2.95) we find for the first subsystem

$$dS_1 = \frac{1}{T_1} dU_1 + \frac{p_1}{T_1} dV_1 + \frac{\xi_1}{T_1} dX_1 - \frac{\varphi^*_1}{T_1} dG_1 \tag{2.134}$$

and similarly for the second subsystem

$$dS_2 = \frac{1}{T_2} dU_2 + \frac{p_2}{T_2} dV_2 + \frac{\xi_2}{T_2} dX_2 - \frac{\varphi^*_2}{T_2} dG_2. \tag{2.135}$$

Substituting these relations into (2.132) and using (2.127)–(2.130), we find

$$\left(\frac{1}{T_1} - \frac{1}{T_2}\right) dU_1 + \left(\frac{p_1}{T_1} - \frac{p_2}{T_2}\right) dV_1 + \left(\frac{\xi_1}{T_1} - \frac{\xi_2}{T_2}\right) dX_1 -$$

$$-\left(\frac{\varphi^*_1}{T_1} - \frac{\varphi^*_2}{T_2}\right) dG_1 = 0. \tag{2.136}$$

Since the differentials dU_1, dV_1, dX_1, and dG_1 are completely independent (according to considerations similar to those developed in Section 2.3 for the independence of dU_1, dV_1, and dX_1), in order that the left-hand side of Eq. (2.136) be identically zero it is necessary that the coefficients of the differentials dU_1, dV_1, dX_1, and dG_1 be separately and independently equal to zero in this equation, i.e., that

$$\frac{1}{T_1} - \frac{1}{T_2} = 0; \tag{2.137}$$

$$\frac{p_1}{T_1} - \frac{p_2}{T_2} = 0, \tag{2.138}$$

$$\frac{\xi_1}{T_1} - \frac{\xi_2}{T_2} = 0, \tag{2.139}$$

and

$$\frac{\varphi^*_1}{T_1} - \frac{\varphi^*_2}{T_2} = 0. \tag{2.140}$$

From the first three relations we find the conditions which we already know,

$$T_1 = T_2, \tag{2.120}$$

$$p_1 = p_2, \tag{2.121}$$

and

$$\xi_1 = \xi_2; \tag{2.122}$$

from Eq. (2.140) we find a new condition,

$$\varphi^*_1 = \varphi^*_2. \tag{2.141}$$

Thus if two phases are in equilibrium then the temperature, pressure, force ξ, and chemical potential of these phases must be equal. Similar methods can be used to give the same result for a system in which three phases coexist.

With regard to the applicability of these results we must make the same remarks as at the end of Section 2.3. In particular it should be stressed that the derivation has not considered certain special properties of the interface of the coexisting phases. This question will be treated in detail below in Section 6.5.

2.5. The Maxwell Equations

We write Eq. (1.28b) in the form

$$du = Tds - pdv - \xi dx. \tag{2.142}$$

For the case v = const this becomes

$$du = Tds - \xi dx. \tag{2.143}$$

Since the internal energy u is a constant function and consequently its differential is a total differential, in agreement with Eq. (1.43) we find from (2.143)

$$\left(\frac{\partial x}{\partial T}\right)_{s,\,v} = -\left(\frac{\partial s}{\partial \xi}\right)_{x,\,v}. \tag{2.144}$$

Similarly, for x = const we find

$$\left(\frac{\partial v}{\partial T}\right)_{s,\,x} = -\left(\frac{\partial s}{\partial p}\right)_{v,\,x}. \tag{2.145}$$

Equation (2.69)

$$di^* = Tds + vdp + xd\xi$$

when p = const takes the form

$$di^* = Tds + xd\xi. \tag{2.146}$$

Since the enthalpy i* is a constant function, in agreement with (1.43), we find

$$\left(\frac{\partial x}{\partial s}\right)_{\xi,\,p} = \left(\frac{\partial T}{\partial \xi}\right)_{s,\,p}. \tag{2.147}$$

Similarly, for ξ = const we find

$$\left(\frac{\partial v}{\partial s}\right)_{p,\,\xi} = \left(\frac{\partial T}{\partial p}\right)_{s,\,\xi}. \tag{2.148}$$

Equation (2.77)

$$df = -pdv - \xi dx - sdT$$

assumes the form

$$df = -\xi dx - sdT \tag{2.149}$$

for v = const.

Since the free energy f is a state function, in agreement with (1.43), we find from (2.149)

$$\left(\frac{\partial x}{\partial s}\right)_{T,\,v}=\left(\frac{\partial T}{\partial \xi}\right)_{x,\,v}. \tag{2.150}$$

Similarly, for x = const we find

$$\left(\frac{\partial v}{\partial s}\right)_{T,\,x}=\left(\frac{\partial T}{\partial p}\right)_{v,\,x}. \tag{2.151}$$

Finally, Eq. (2.85)

$$d\varphi^* = v\,dp + x\,d\xi - s\,dT$$

assumes the form

$$d\varphi^* = x\,d\xi - s\,dT \tag{2.152}$$

for p = const.

Since the chemical potential Φ^* is a state function, we find in agreement with (1.43)

$$\left(\frac{\partial x}{\partial T}\right)_{\xi,\,p}=-\left(\frac{\partial s}{\partial \xi}\right)_{T,\,p}. \tag{2.153}$$

Similarly for ξ = const we find

$$\left(\frac{\partial v}{\partial T}\right)_{p,\,\xi}=-\left(\frac{\partial s}{\partial p}\right)_{T,\,\xi}. \tag{2.154}$$

The differential equations (2.144), (2.145), (2.147), (2.148), (2.150), (2.151), (2.153), and (2.154) which establish relations between thermal (T and s) and nonthermal (ξ and x or p and v) parameters are called the Maxwell equations.

While we have found the equations for the weight-specific quantities s, v, and x, the Maxwell equations have the same form for the total quantities S, V, and X:

$$\left(\frac{\partial X}{\partial T}\right)_{S,v}=-\left(\frac{\partial S}{\partial \xi}\right)_{X,v}, \tag{2.144a}$$

$$\left(\frac{\partial V}{\partial T}\right)_{S,x}=-\left(\frac{\partial S}{\partial p}\right)_{V,x}, \tag{2.145a}$$

$$\left(\frac{\partial X}{\partial S}\right)_{\xi,\,p} = \left(\frac{\partial T}{\partial \xi}\right)_{S,\,p}, \qquad (2.147a)$$

$$\left(\frac{\partial V}{\partial S}\right)_{p,\,\xi} = \left(\frac{\partial T}{\partial p}\right)_{S,\,\xi}, \qquad (2.148a)$$

$$\left(\frac{\partial X}{\partial S}\right)_{T,V} = \left(\frac{\partial T}{\partial \xi}\right)_{X,V}, \qquad (2.150a)$$

$$\left(\frac{\partial V}{\partial S}\right)_{T,\,x} = \left(\frac{\partial T}{\partial p}\right)_{V,x}, \qquad (2.151a)$$

$$\left(\frac{\partial X}{\partial T}\right)_{\xi,\,p} = -\left(\frac{\partial S}{\partial \xi}\right)_{T,\,p}, \qquad (2.153a)$$

$$\left(\frac{\partial V}{\partial T}\right)_{p,\,\xi} = -\left(\frac{\partial S}{\partial p}\right)_{T,\,\xi}. \qquad (2.154a)$$

These important equations will be used many times in what follows.

For instance, the Maxwell equations allow one to calculate the difference between the heat capacities at constant generalized force c_ξ and at constant generalized coordinate c_x. In agreement with Eq. (1.49), we can write

$$\left(\frac{\partial s}{\partial T}\right)_\xi = \left(\frac{\partial s}{\partial T}\right)_x + \left(\frac{\partial s}{\partial x}\right)_T \left(\frac{\partial x}{\partial T}\right)_\xi. \qquad (2.155)$$

Replacing $(\partial s/\partial x)_T$ in this relation by the Maxwell equation (2.150)

$$\left(\frac{\partial s}{\partial x}\right)_T = \left(\frac{\partial \xi}{\partial T}\right)_x$$

and taking (1.37) and (1.38) into account, we find

$$c_\xi - c_x = T\left(\frac{\partial \xi}{\partial T}\right)_x \left(\frac{\partial x}{\partial T}\right)_\xi. \qquad (2.156)$$

For "simple" systems for which we can set $\xi = p$ and $x = v$ we find the well-known relation

$$c_p - c_v = T\left(\frac{\partial p}{\partial T}\right)_v \left(\frac{\partial v}{\partial T}\right)_p. \qquad (2.157)$$

CHAPTER 3

Magnets

3.1. Introduction

As is well known, the most important property of the magnetic field is the vector-valued field strength H.

Magnetic properties are present to some degree in all materials; therefore in treating the magnetic properties of materials we will use the general term magnets. A magnet is a medium which is able to affect the external magnetic field (to strengthen or weaken this field). The effect of a magnet on the external magnetic field can be represented as follows in simplified form. The magnet can be regarded as a set of a huge number of elementary magnets — magnetic dipoles,† oriented more or less randomly in the absence of a field. Within a magnet there are many such magnetic dipoles but because of their random arrangement their magnetic fields cancel each other. When such a magnet is placed in a magnetic field it becomes magnetized; the magnetic dipoles become oriented along the field (or opposite to the field), and the inherent magnetic fields of these dipoles add together making a kind of macroscopic magnet whose field strengthens (or weakens) the external magnetic field. Thus in the magnetized state a magnet makes an additional contribution H' to the magnetic field strength which sums with the initial strength H. The sum of H and H' which thus characterizes the total magnetization of the medium is called the magnetic induction field‡ :

$$B = H + H'. \tag{3.1}$$

---

† We will show below that this idea is not applicable for one type of magnet (a diamagnet).

‡ Here and in Chapters 4 and 5 we use Gaussian units.

47

The degree to which a medium is magnetized (the magnetization) is also characterized by the magnetization vector,[†] which is defined as the magnetic moment per unit volume of the magnet

$$J = \frac{\sum\limits_{i=1}^{n} j_i}{V},$$ (3.2)

where $\sum\limits_{i=1}^{n} j_i$ is the geometric sum of the molecular magnetic moments (j_i) in a magnet which occupies the volume V.

It is to be understood that J is an extensive quantity. We will show below that B and J are uniquely related.

Evidently J depends on H; the stronger the external magnetic field (the higher H) the more the magnetic dipoles are oriented along the field and the greater the magnetization J. As we know from physics, the magnetization is related to the strength of the external magnetic field by

$$J = \varkappa H,$$ (3.3)

where the coefficient $\varkappa$ is called the magnetization coefficient, or the magnetic susceptibility. The magnetic susceptibility is an individual property of each magnetic material. It varies with temperature; for example, for a paramagnet $\varkappa$ is lower the higher the temperature since with increasing temperature the random thermal motion is intensified and inhibits the external magnetic field from "ordering" the orientations of the magnetic dipoles. As for the dependence of $\varkappa$ on the magnitude H of the external magnetic field, $\varkappa$ depends on H for some magnets but not for others.

It is thus clear that the additional magnetic field strength H' created by a magnet in the presence of an external magnetic field H is uniquely related to the magnetization J.

We recall from general physics that the relation between H' and J has the following form:

$$H' = 4\pi J.$$ (3.4)

[†]Throughout this chapter we consider isotropic uniformly magnetized bodies. For more complex cases see L. D. Landau and E. M. Lifshits, Electrodynamics of Continuous Media, Fizmatgiz (1959).

Using this relation, we find from (3.1)

$$B = H + 4\pi J. \tag{3.5}$$

Replacing J in this equation using Eq. (3.3), we find

$$B = (1 + 4\pi\varkappa) H \tag{3.6}$$

or

$$B = \mu H, \tag{3.7}$$

where

$$\mu = 1 + 4\pi\varkappa \tag{3.8}$$

is called the magnetic permeability of the medium.

It should be stressed that the quantity J which enters the relations given above is, as we see from (3.2), the magnetization per unit volume of the magnet. Correspondingly the magnetic susceptibility $\varkappa$ defined by Eq. (3.3) refers to a unit volume of the magnet. Evidently the specific magnetization j per unit weight of the magnet is related to J by

$$j = Jv, \tag{3.9}$$

where v is the specific volume of the magnet. For the specific magnetization, we write Eq. (3.3) in the form

$$j = \chi H, \tag{3.10}$$

where χ is the so-called weight-specific magnetic susceptibility.[†]

It follows from (3.3), (3.9), and (3.10) that

$$\chi = \varkappa v. \tag{3.11}$$

Finally it is clear that the total magnetic moment of the whole magnet is given by

$$\mathfrak{J} = JV = jG, \tag{3.12}$$

[†] It is clear that Eqs. (3.3) and (3.10) can be regarded as equations of state for a magnet.

where V and G are, respectively, volume and weight of the magnet.

Below we will use mainly the specific values j and χ.

The values of χ (or $\varkappa$) can be either positive or negative. If $\chi > 0$ the direction of the magnetization vector J coincides with the direction of the external magnetic field vector H, i.e., the total magnetic field of the oriented magnetic dipoles runs in the same direction as the external magnetic field; consequently B > H in agreement with (3.1). If $\chi < 0$ then the directions of the vectors J and H are opposite, i.e., the total field of the magnetic dipoles subtracts from H; in this case B < H in agreement with (3.1).

Depending on the sign of χ and the nature of the relation $\chi = f(\text{H})$, magnets can be divided into two main classes, diamagnets and paramagnets.

Magnets in which the magnetic susceptibility is negative ($\chi < 0$) and independent of the external magnetic field strength H are called diamagnets. Since $\chi < 0$, a diamagnet weakens the external magnetic field (B < H). Examples of diamagnets include such materials as carbon, mercury, gold, silver, copper, zinc, phosphorus, antimony, bismuth, and many chemical compounds (including almost all organic compounds), nitrogen, carbonic acid, ammonia, chlorine, hydrogen, all of the inert gases, water, oil, wood, marble, glass, etc.

In agreement with Langevin's theory, diamagnetism arises from procession of the electronic orbits of the atoms about the external magnetic field direction. As a result of this precession the atom acquires a magnetic moment opposite to the externally applied magnetic field.[†] Consequently the magnetization vector of the diamagnet as a whole lies opposite to the field. (As we see, diamagnetism cannot be explained on the basis of ideas about elementary magnets given at the beginning of this section.)

It should be stressed that for the majority of diamagnetic materials the magnetic susceptibility χ is very small and almost independent of temperature; therefore for most diamagnets we can assume that $(\partial\chi/\partial T)_H = 0$.

[†]This effect occurs in any material including the paramagnets and ferromagnets described above; however in these materials the effect coincides with the substantially stronger paramagnetic or ferromagnetic effects.

The fact that χ is independent of temperature for a diamagnet can be explained by the fact that the intensity of the thermal motion hardly affects the precession of the electron orbits in a magnetic field.

A substantial temperature dependence of χ only occurs for a small number of diamagnetic materials having anomalously large values of χ. Such diamagnetic materials include, for example, bismuth and thallium. In thallium χ is approximately 20% higher at 14°K than at room temperature.[†]

At low temperatures (of the order of several degrees Kelvin) χ varies somewhat with H in some metals. Such metals include zinc, tin, bismuth (single crystal), etc. For the majority of diamagnetic materials, as previously noted, the magnetic susceptibility is independent of the external magnetic field strength over a broad temperature range.

Magnets in which the magnetic susceptibility is positive ($\chi > 0$) are called paramagnets. Since $\chi > 0$, a paramagnet strengthens the external magnetic field (B > H). For a number of paramagnets χ does not depend on the magnitude of the external magnetic field strength H. Such paramagnets include molecular oxygen, platinum, the rare-earth elements, iron-group element salts, alkali metals, etc.

Each atom (or molecule) of a paramagnet has an uncompensated magnetic moment and consequently behaves as a magnetic dipole. As previously noted at the beginning of this section, in the absence of an external magnetic field the dipoles are oriented in a disordered fashion and their total magnetic moment is zero. When an external magnetic field is applied the orientation of the magnetic dipoles becomes better the higher the magnetic field strength. This orientation also results in a nonzero total magnetic moment lying along the field (i.e., $\chi > 0$). An increase in the temperature disorders this orientation due to an increase in the intensity of the random thermal motion (χ drops with increasing temperature).

For most paramagnets the magnetic susceptibility χ varies strongly with temperature.

The temperature dependence of χ for most paramagnets at not too low temperatures is quite well described by a relation called the Curie law

$$\chi = \frac{A}{T},$$

(3.13)

where A is a constant (the so-called Curie constant) which is different for different materials.

[†]Note that the nature of the diamagnetism in these metals is somewhat different from the scheme developed above.

This law was experimentally established in 1906 by P. Curie and then given a theoretical basis by P. Langevin. It is rigorously valid only for a limited group of paramagnetic materials (certain gases, platinum, palladium, rare-earth elements and their salts, etc.). The great majority of paramagnetic materials obey the Curie-Weiss law

$$\chi = \frac{C}{T + \Delta}, \tag{3.14}$$

where Δ is a constant which depends on the nature of the material (it can be either positive or negative).

It should be noted that the Curie law remains valid at temperatures close to absolute zero. In agreement with the Nernst law

$$\lim_{T \to 0^\circ \text{K}} \left(\frac{\partial j}{\partial T} \right)_H = 0 \tag{3.15}$$

and consequently, using (3.10),

$$\lim_{T \to 0^\circ \text{K}} \left(\frac{\partial \chi}{\partial T} \right)_H = 0. \tag{3.16}$$

This law does not follow from Eq. (3.13).

In some paramagnetic metals (solid and liquid alkali metals, and alkaline-earth metals) whose paramagnetism arises from the spin magnetic moments of the conduction electrons, the magnetic susceptibility χ is almost independent of temperature.

It should be noted that for large values of the magnetic field strength χ becomes dependent on H in the sense that the paramagnet saturates for such values of H; all of the magnetic dipoles are oriented strictly along the field, and a further increase in H does not increase the magnetization j of the paramagnet.

Ferromagnets form a special class of paramagnets. In ferromagnetic materials the magnetic susceptibility ($\chi > 0$) varies strongly with a change in the strength H of the external magnetic field. Since $\chi > 0$, a ferromagnet enhances the magnetic field. Iron, nickel, and cobalt are examples of ferromagnetic materials. The essential difference between ferromagnets and other paramagnetic materials is that in ferromagnetic materials the additional field strength H' is many orders of magnitude larger than in ordinary paramagnetic materials (in other words, the magnetic susceptibility of a ferromagnetic is many times greater than in an ordinary paramagnet). This can be explained as follows. Instead of individual molecular magnetic dipoles, ferromagnets have substantially larger elementary groups — the so-called domains (regions of spontaneous magnetization consisting of associations of magnetic dipoles). The domains are easily oriented along the field and, what is very important, remain oriented after the magnetic field is removed. Thus a ferromagnet remains magnetized even after the external field is removed (while in diamagnets and paramagnets the magnetization vanishes in zero external magnetic field).

In contrast to diamagnets and paramagnets the relation $j = f(H)$ is nonlinear for ferromagnets, i.e., $\chi = (\partial j/\partial H)_T$†changes with H. The magnetization of a ferromagnet increases with increasing H, reaching saturation ($j = j_s$) for some value $H = H_s$. The nature of the relation $j = f(H)$ for a ferromagnet is shown in Fig. 3.1.

Residual magnetization can exist in ferromagnets at temperatures not exceeding the so-called Curie point temperature (θ). For iron the Curie point is 770°C, for cobalt it is 1150°C, and for nickel it is 360°C. In approaching the Curie point the residual magnetization of a ferromagnet decreases, and it becomes zero at the Curie point. At the Curie point there is a second-order phase transition in a ferromagnet (from the ferromagnetic to the paramagnetic state). With $T > \theta$ the predominant orientation of the magnetic fields in the domains vanishes and the ferromagnet behaves as an ordinary paramagnet at these temperatures.

Figure 3.2 illustrates the nature of the temperature dependence of χ for a ferromagnet (this graph shows the temperature dependence of the magnetic susceptibility $\mu = 1 + 4\pi\chi = 1 + (4\pi\chi)/v$ for iron with H = 0). As we see from this graph, the magnetic susceptibility of a ferromagnet increases in approaching the Curie point with H = const reaching a maximum near the Curie point, while in the immediate vicinity of the Curie point it sharply decreases — the so-called Hopkinson effect (this effect is observed only in weak magnetic fields). This maximum arises from a substantial decrease in the magnetic anisotropy of the ferromagnet near the Curie point such that a ferromagnet becomes "easier" to magnetize while χ decreases with a further approach to the Curie point because the spontaneous magnetization of the ferromagnet vanishes near $T = \theta$. With $T > \theta$, χ continues to decrease with increasing temperature, and the relation between χ and T in this region is described by the Curie–Weiss law (3.14); in this case $\Delta = -\theta$ in Eq. (3.14).

It should further be stressed that the magnetic susceptibility decreases with in-

---

†It is to be understood that for a nonlinear relation $j(H)$ instead of the magnetic susceptibility $\chi = j/H$ defined by Eq. (3.3) or (3.10) we must introduce the differential magnetic susceptibility $\chi = (\partial j/\partial H)_T$.

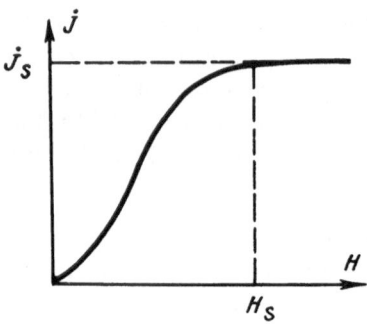

Fig. 3.1

creasing temperature in paramagnets, i.e., $(\partial \chi / \partial T)_H < 0$ and consequently $(\partial j / \partial T)_H < 0$; it is to be understood that this law is not valid for ferromagnets in the region where the Hopkinson effect appears. The derivative $(\partial \chi / \partial T)_H$ is positive to the left of the maximum in $\chi(T)$.

In completing the list of basic kinds of magnets we should mention the so-called antiferromagnets. Antiferromagnets are a special class of paramagnets; their magnetic susceptibilities χ are positive, but in contrast to other paramagnetic materials, for an antiferromagnet χ increases with increasing temperature. This increase continues to a certain temperature which is called the antiferromagnetic Curie point (θ_{af}). At this point a second-order phase transformation occurs in an antiferromagnet and above θ_{af} the magnetic susceptibility of an antiferromagnet varies by the Curie–Weiss law usually found for paramagnetic materials; in this case $\Delta = -\theta_{af}$ in Eq. (3.14). The nature of the temperature dependence of χ for an antiferromagnetic material is shown in Fig. 3.3. It should also be noted that with $T < \theta_{af}$, χ changes substantially with changing H.

Antiferromagnetic materials include a number of transition elements and their compounds and alloys (such as Cr, MnO, MnTe, MnSe, $FeCl_2$, FeO, $CrCl_2$, etc.).

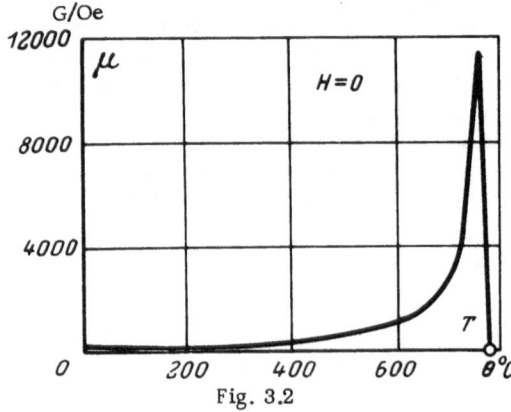

Fig. 3.2

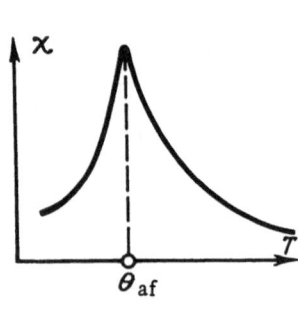

Fig. 3.3

It can be seen from (3.8) that in diamagnets ($\chi < 0$) the magnetic susceptibility is such that $\mu < 1$ while in paramagnets ($\chi > 0$), including ferromagnets, $\mu > 1$.

The nature of the relations $j = f(H)$, $B = f(H)$, $\chi = f(H)$, and $\chi = f(T)$ for various kinds of magnets is shown in Fig. 3.4.

One important fact should be noted. The magnetic field strength in a magnet will equal the external uniform magnetic field strength only for a long magnet of cylindrical shape in a longitudinal magnetic field. In all other cases the presence of the magnet distorts the external field and the corresponding relations [Eq. (3.3), etc.] should contain the value H_{int} of the magnetic field within the magnet, which differs from the strength H of the external magnetic field. For paramagnets and ferromagnets, $H_{int} < H$ and for diamagnetic materials $H_{int} > H$. The values of H and H_{int} are related by

$$H_{int} = H - 4\pi nJ,$$

where n is the demagnetization coefficient (demagnetizing factor), and using (3.3) it follows that

$$H_{int} = \frac{H}{1 + 4\pi n\varkappa}.$$

It is evident from this relation that for a magnet of arbitrary shape this effect can only be neglected when $\varkappa$ is small relative to unity (paramagnets and diamagnets). Neglect of this effect would lead to serious errors for ferromagnets.

The value of n is 1/3 for a sphere, tends to 1 for a planar disc and 1/2 for a long cylinder in a transverse field, and is zero for a long cylinder in a longitudinal field.

The corresponding equations given in Chapter 4 will relate similarly to a long cylindrical dielectric in a uniform longitudinal electric field. In all other cases we must introduce coefficients into these relations which are "dielectric analogs" of the demagnetization coefficient.

3.2. Basic Thermodynamic Relations for Magnets

As we know from physics, the elementary work accomplished by a magnetic field of strength H in increasing the magnetization of a magnet from J to J + dJ is[†]

$$dL^* = -HdJ \tag{3.17}$$

[†]Since, as noted above, J and H are vector quantities, it must be kept in mind that HdJ is the scalar product of the vectors H and dJ (in other words, dJ is the projection of the vector dJ along the direction of the field H).

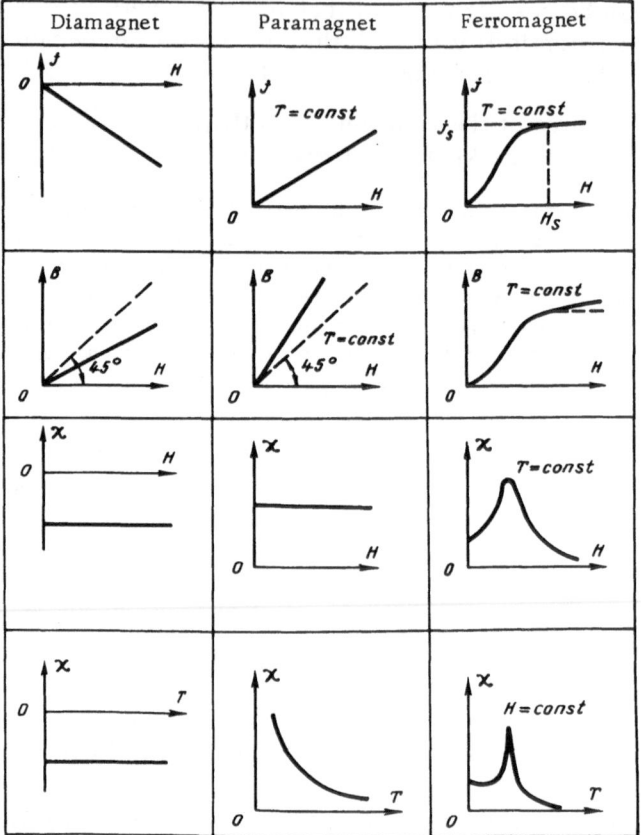

Fig. 3.4

or per unit weight of the magnet

$$dl^* = -Hdj.$$ (3.18)

The minus sign shows that work is done on the magnet with increasing magnetization.

Using Eqs. (3.5) and (3.7), Eq. (3.17) may be written in the form

$$dL^* = \frac{\mu - 1}{4\pi\mu} HdB = \frac{\mu - 1}{4\pi} HdH .$$ (3.17a)

In addition to the work done in magnetizing the magnet, as we know from electrodynamics there is work done in increasing the magnetic field strength from H to H + dH in the absence of a magnet (in vacuum). This elementary work turns out to be (per unit volume)

$$dL\text{vac} = -\frac{1}{4\pi} HdH,$$ (3.19)

and consequently in order to create a magnetic field of strength H in a given volume V in the absence of a magnet we must do work equal to

$$L_{\text{vac}} = -\frac{H^2}{8\pi}.$$ (3.20)

Evidently L_{vac} is the energy of the magnetic field in the absence of a magnet (in vacuum).

In view of these considerations the so-called total work of the magnetic field equals the sum of the work done in increasing the magnetic field strength in vacuum and that done magnetizing the magnet and is (per unit volume).

$$dL\text{tot} = -\frac{1}{4\pi} HdH - HdJ = -\frac{HdB}{4\pi} = -\frac{\mu HdH}{4\pi}.$$ (3.21)

The thermodynamic relations for magnets can be written either taking L_{vac} into account or without considering this quantity; both ways are equally valid since L_{vac} does not contribute to the thermodynamic functions. Here we do not use this formulation of the work. We use only the magnitude of the work of magnetization given by Eqs. (3.17) and (3.18).

Thus in the system under consideration (a magnet in a magnetic field) the generalized force ξ is the magnetic field strength H and the generalized coordinate x is the magnetization j of the magnet.

The generalized equation of the first and second laws of thermodynamics [Eq. (1.28a)] for a system in a magnetic field is written as follows:

$$Tds = du + pdv - Hdj$$ (3.22)

(here and below all of the relations are given for weight-specific values of the thermodynamic quantities).

In agreement with (2.66) the enthalpy i* for a system in a magnetic field is

$$i^* = u + pv - Hj$$ (3.23)

or, what is the same,

$$i^* = i - Hj.$$ (3.24)

Taking (3.23) into account, Eq. (3.22) can be written as follows:

$$Tds = di^* - vdp + jdH.$$ (3.25)

It further follows from (2.53) that the specific isobaric–isothermal potential (i.e., the chemical potential) of such a system is

$$\varphi^* = u + p\mathbf{v} - Hj - Ts \tag{3.26}$$

or

$$\varphi^* = i^* - Ts. \tag{3.27}$$

In agreement with (2.144), (2.147), (2.150), and (2.153) the Maxwell equations for a system in a magnetic field are written as follows:

$$\left(\frac{\partial j}{\partial T}\right)_{s,\,v} = \left(\frac{\partial s}{\partial H}\right)_{j,\,v}, \tag{3.28}$$

$$\left(\frac{\partial j}{\partial s}\right)_{H,\,p} = -\left(\frac{\partial T}{\partial H}\right)_{s,\,p}, \tag{3.29}$$

$$\left(\frac{\partial j}{\partial s}\right)_{T,\,v} = -\left(\frac{\partial T}{\partial H}\right)_{j,\,v}, \tag{3.30}$$

$$\left(\frac{\partial j}{\partial T}\right)_{H,\,p} = \left(\frac{\partial s}{\partial H}\right)_{T,\,p}. \tag{3.31}$$

It is not hard to find relations between the caloric properties of a magnet, its internal energy and enthalpy, and j and H from (3.22) and (3.25) using the Maxwell equations. (It should be noted that the roles of j and H for magnets are to some degree analogous to the roles of v and p, respectively, for "ordinary" thermodynamic systems.)

We find from (3.22) that

$$\left(\frac{\partial u}{\partial j}\right)_{T,\,v} = T\left(\frac{\partial s}{\partial j}\right)_{T,\,v} + H, \tag{3.32}$$

from which

$$\left(\frac{\partial u}{\partial j}\right)_{T,\,v} = H - T\left(\frac{\partial H}{\partial T}\right)_{j,\,v} \tag{3.33}$$

using (3.30). The physical meaning of the derivative $(\partial H/\partial T)_j$

is clear: it shows how much it is necessary to increase H with increasing temperature of the magnet to hold the specific magnetization j unchanged (regardless of the growth of temperature, which changes the magnetization as noted previously).

Having substituted H into Eq. (3.33) in agreement with (3.10), we find

$$\left(\frac{\partial u}{\partial j}\right)_{T,\,v} = \frac{j}{\chi}\left[1 + \frac{T}{\chi}\left(\frac{\partial \chi}{\partial T}\right)_{j,\,v}\right]. \tag{3.34}$$

For diamagnets in which, as noted above, the magnetic susceptibility χ does not depend on the temperature, this equation, assumes the form

$$\left(\frac{\partial u}{\partial j}\right)_{T,\,v} = \frac{j}{\chi}. \tag{3.35}$$

Integrating, we find the relation which allows us to determine how much the internal energy of a diamagnet changes as its magnetization changes from 0 to j,

$$u_2(j, T, v) - u_1(j = 0, T, v) = \frac{j^2}{2\chi} \tag{3.36}$$

or, what is the same, as the magnetic field goes from 0 to H,

$$u_2(H, T, v) - u_1(H = 0, T, v) = \frac{\chi H^2}{2} \tag{3.37}$$

or, finally,

$$u_2(H, T, v) - u_1(H = 0, T, v) = \frac{jH}{2}. \tag{3.38}$$

Since $\chi < 0$ in diamagnets it follows from (3.37) that the internal energy of the diamagnet decreases along an isotherm (and with v = const) with increasing field strength by a parabolic law. Since $\chi \neq f(T)$ for a diamagnet, as we see from (3.37) for a given value of H the internal energy of a diamagnet decreases with increasing H by the same amount for different isotherms.

As for paramagnets and ferromagnets, as noted above we

have $(\partial\chi/\partial T)_{H,\,p} < 0$ — the magnetic susceptibility decreases with increasing temperature.[†] Because of this it follows from (3.34) that $(\partial u/\partial j)_{T,\,v}$ in general can be either positive or negative for these types of magnet. Evidently if $|(\partial\chi/\partial T)_{j,\,v}| > (\chi/T)$ then $(\partial u/\partial j)_{T,\,v} < 0$. If $|(\partial\chi/\partial T)_{j,\,v}| < (\chi/T)$, then $(\partial u/\partial j)_{T,\,v} > 0$.

The thermodynamic relations for magnets can be written either taking L_{vac} into account or without considering this quantity; both ways are equally valid since L_{vac} does not contribute to the thermodynamic functions. Here we do not use this formulation of the work. We use only the magnitude of the work of magnetization given by Eqs. (3.17) and (3.18).

We consider one particular variety of paramagnet.

It follows from (3.10) and (3.13) that for paramagnets which obey the Curie law we can write

$$H = \frac{jT}{A} \tag{3.39}$$

and, consequently,

$$\left(\frac{\partial H}{\partial T}\right)_{j,\,v} = \frac{H}{T}. \tag{3.40}$$

Using this relation, we find from (3.33) that

$$\left(\frac{\partial u}{\partial j}\right)_{T,\,v} = 0 \tag{3.41}$$

is the internal energy of a paramagnet which obeys the Curie law and does not depend on the magnetization. Such paramagnets are called ideal paramagnets.[‡]

It is also evident from (3.41) and (3.10) that for an ideal paramagnet,

$$\left(\frac{\partial u}{\partial H}\right)_{T,\,v} = 0. \tag{3.42}$$

We find for the enthalpy of a magnet from (3.25)

$$\left(\frac{\partial i^*}{\partial H}\right)_{T,\,\nu} = T\left(\frac{\partial s}{\partial H}\right)_{T,\,\nu} - j, \tag{3.43}$$

[†]Excluding the left-hand branch of the $\chi(T)$ curve in the region of the Hopkinson effect.

[‡]There is a definite thermodynamic analogy with the idea of an ideal gas in which the internal energy is always independent of the generalized coordinate, the volume:

$$\left(\frac{\partial u}{\partial v}\right)_T = 0.$$

whence using (3.31) we find

$$\left(\frac{\partial i^*}{\partial H}\right)_{T,p} = T\left(\frac{\partial j}{\partial T}\right)_{H,p} - j.$$ (3.44)

Using Eq. (3.10), Eq. (3.44) can be transformed to

$$\left(\frac{\partial i^*}{\partial H}\right)_{T,p} = H\left[T\left(\frac{\partial \chi}{\partial T}\right)_{H,p} - \chi\right].$$ (3.45)

For diamagnets in which the magnetic susceptibility is independent of the temperature we find

$$\left(\frac{\partial i^*}{\partial H}\right)_{T,p} = -\chi H.$$ (3.46)

Integrating this relation (and using the fact that χ does not change with H for a diamagnet), we find

$$i^*(H,T,p) - i(H=0,T,p) = -\frac{\chi H^2}{2}.$$ (3.47)

This equation shows that (because $\chi < 0$ for diamagnets) the enthalpy of a diamagnet increases as the magnetic field strength increases from 0 to H.

Equation (3.47) can also be obtained by substituting u from Eq. (3.37) into Eq. (3.23) using Eq. (3.10). Since $\chi \neq f(T)$ for a diamagnet, for a given value of H the increase of the enthalpy with increasing H will be the same for different isotherms.

It follows from (3.47) that the enthalpy of a diamagnet increases by a parabolic law with increasing field strength. We note that the internal energy of a diamagnet decreases by the same law. There is nothing remarkable in this — with increasing H the internal energy u of a diamagnet decreases by $\chi H^2/2$ but since it is necessary in agreement with (3.23) to add the quantity (pv−Hj) = (pv−χH^2) to u to calculate the enthalpy, then despite the decrease in u the enthalpy i^* increases by an amount $-\chi H^2/2$.

The nature of the changes in the internal energy and enthalpy of a diamagnet with a change in the magnetic field strength H is shown schematically in Fig. 3.5.

Since $(\partial \chi / \partial T)_H < 0$ for paramagnets and ferromagnets, we see from (3.45) that for these types of magnet

$$\left(\frac{\partial i^*}{\partial H}\right)_{T,p} < 0, \tag{3.48}$$

i.e., under isothermal conditions the enthalpy i^* of a magnet decreases with increasing external magnetic field strength; this fact is closely related to the magnetocaloric effect discussed above.

For an ideal paramagnet (which obeys the Curie law) as we see from (3.13)

$$\left(\frac{\partial \chi}{\partial T}\right)_{H,p} = -\frac{\chi}{T}. \tag{3.49}$$

Substituting this value of $(\partial \chi / \partial T)_{H,p}$ into Eq. (3.45), we find

$$\left(\frac{\partial i^*}{\partial H}\right)_{T,p} = -2j = -2\chi H; \tag{3.50}$$

in contrast to the internal energy, which in agreement with Eq. (3.42) does not depend on H, the enthalpy of an ideal paramagnet decreases with increasing magnetic field strength in proportion to H.

It is also not hard to show that

$$\left(\frac{\partial i^*}{\partial j}\right)_{T,p} = -2H. \tag{3.51}$$

for an ideal paramagnet.

In speaking of the enthalpy of a magnet we should make one essential remark. At first glance we would think that it would follow from Eq. (3.24)

$$i^* = i - Hj$$

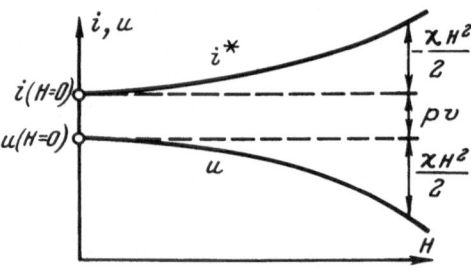

Fig. 3.5

that the enthalpy i^* differs by $-Hj$ from the enthalpy of a magnet in the absence of a magnetic field (H = 0) since i also is the enthalpy of a magnet at H = 0. However, this is not really so. The quantity i = u + pv in turn changes with changing H because, as shown above, as H changes there is a change in the internal energy of the magnet: it follows from Eq. (3.33) that $(\partial u/\partial j)_T$ and $(\partial u/\partial H)_T$ in general are non-zero.

Equation (3.23) can be put in the form

$$i^* = u_{H=0} + \Delta u_H + pv - Hj, \tag{3.52}$$

where $u_{H=0}$ is the internal energy of a magnet in the absence of a magnetic field and Δu_H is the change in the internal energy of a magnet with an increase of the magnetic field strength from 0 to H. It is to be understood that

$$\Delta u_H = \int_0^H \left(\frac{\partial u}{\partial H} \right)_{T,\,p} dH. \tag{3.53}$$

Equation (3.52) can be written in the form

$$i^* = i_{H=0} + \Delta u_H - Hj; \tag{3.54}$$

here $i_{H=0} = u_{H=0} + pv$ is the part of the total enthalpy of the magnet which is independent of H.

Strictly speaking with a change in the magnetic field strength not only will the internal energy of the magnet change but also its specific volume — the so-called magnetostriction effect (see Section 3.5). If this effect is taken into account it is evident that the quantity v which enters (3.52) is the specific volume of a magnet in a field of strength H, which differs somewhat from the specific volume with H = 0. It follows from this that Eq. (3.52) can be written in the form

$$i^* = u_{H=0} + \Delta u_H + pv_{H=0} + p\Delta v_H - Hj, \tag{3.52a}$$

where

$$\Delta v_H = v_H - v_{H=0} = \int_0^H \left(\frac{\partial v}{\partial H} \right)_{T,\,p} dH.$$

In view of this, Eq. (3.54) assumes the following form:

$$i^* = i_{H=0} + \Delta u_H + p\Delta v_H - Hj = i_{H=0} + \Delta i_H - Hj. \tag{3.54a}$$

For ideal paramagnets, in which the internal energy does not change with a change of H in agreement with (3.42) and consequently $\Delta u_H = 0$, we find from (3.54) using (3.10)

$$i^* = i_{H=0} - \chi H^2. \tag{3.55}$$

3.3. The Heat Capacities of a Magnet

By analogy with the isobaric and isochoric heat capacities, for a magnet we introduce the idea of heat capacities at constant magnetic field strength c_H and constant magnetization c_j. If these heat capacities are calculated for a constant external pressure of the medium in which the magnet is placed then they are denoted, respectively, by $c_{H,p}$ and $c_{j,p}$. Here

$$c_{H,p} = T\left(\frac{\partial s}{\partial T}\right)_{H,p} \quad \text{and} \quad c_{j,p} = T\left(\frac{\partial s}{\partial T}\right)_{j,p}. \tag{3.56}$$

From the relations

$$\left(\frac{\partial s}{\partial T}\right)_{H,p} = \left(\frac{\partial s}{\partial T}\right)_{j,p} + \left(\frac{\partial s}{\partial j}\right)_{T,p}\left(\frac{\partial j}{\partial T}\right)_{H,p}, \tag{3.57}$$

using the fact that

$$\left(\frac{\partial s}{\partial j}\right)_{T,p} = \left(\frac{\partial s}{\partial H}\right)_{T,p}\left(\frac{\partial H}{\partial j}\right)_{T,p} \tag{3.58}$$

and the Maxwell equation (3.31), we find

$$c_{H,p} - c_{j,p} = -T\left(\frac{\partial j}{\partial T}\right)_{H,p}^2\left(\frac{\partial H}{\partial j}\right)_{T,p}, \tag{3.59}$$

a relation which connects $c_{H,p}$ and $c_{j,p}$ similar to the known thermodynamic relation between c_p and c_v

$$c_p - c_v = -T\left(\frac{\partial v}{\partial T}\right)_p^2\left(\frac{\partial p}{\partial v}\right)_T. \tag{3.60}$$

Furthermore in agreement with (1.48) we can write

$$\left(\frac{\partial j}{\partial T}\right)_H\left(\frac{\partial T}{\partial H}\right)_j\left(\frac{\partial H}{\partial j}\right)_T = -1. \tag{3.61}$$

Replacing the partial derivatives in Eq. (3.59) using this relation we find[†]

$$c_{H,p} - c_{j,p} = -T\left(\frac{\partial j}{\partial T}\right)_{H,p}\left(\frac{\partial H}{\partial T}\right)_{j,p} \tag{3.62}$$

[†]In fact the equation for the difference $c_{H,p} - c_{j,p}$ can be obtained another way by using Eq. (2.156) formulated in Chapter 2.

and

$$c_{H,p} - c_{j,p} = T \left(\frac{\partial H}{\partial T} \right)_{j,p}^{2} \left(\frac{\partial j}{\partial H} \right)_{T,p}.$$ (3.63)

Since in agreement with (3.10) we have

$$j = \chi H,$$

the partial derivatives which enter Eqs. (3.59), (3.62), and (3.63) can be represented as

$$\left(\frac{\partial j}{\partial T} \right)_{H,p} = H \left(\frac{\partial \chi}{\partial T} \right)_{H,p},$$ (3.64)

$$\left(\frac{\partial j}{\partial H} \right)_{T,p} = \chi + H \left(\frac{\partial \chi}{\partial H} \right)_{T,p},$$ (3.65)

and

$$\left(\frac{\partial H}{\partial T} \right)_{j,p} = - \frac{H}{\chi} \left(\frac{\partial \chi}{\partial T} \right)_{j,p}.$$ (3.66)

Using these relations we can elucidate the values of $c_{H,p} - c_{j,p}$ for different types of magnets.

Diamagnets have the following properties:

$$\chi < 0, \quad \left(\frac{\partial \chi}{\partial T} \right)_{H} = \left(\frac{\partial \chi}{\partial T} \right)_{j} = 0, \text{ and } \left(\frac{\partial \chi}{\partial H} \right)_{T} = 0.$$ (3.67)

Consequently for diamagnets

$$\left(\frac{\partial j}{\partial T} \right)_{H} = 0, \quad \left(\frac{\partial j}{\partial H} \right)_{T} < 0, \text{ and } \left(\frac{\partial H}{\partial T} \right)_{j} = 0.$$ (3.68)

Using these relations we find from (3.59), (3.62), or (3.63) that for diamagnets

$$c_{H,p} - c_{j,p} = 0,$$ (3.69)

i.e., the heat capacities are the same for diamagnets.

Paramagnets and ferromagnets have the following properties:

$$0 < \chi \quad \text{and} \quad \left(\frac{\partial \chi}{\partial T} \right)_{H} < 0.$$ (3.70)

With regard to $(\partial\chi/\partial H)_T$, for paramagnets

$$\left(\frac{\partial\chi}{\partial H}\right)_T = 0,$$

(3.71)

while for ferromagnets

$$\left(\frac{\partial\chi}{\partial H}\right)_T \neq 0,$$

(3.72)

so that in agreement with Eq. (3.65) for a paramagnet

$$\left(\frac{\partial j}{\partial H}\right)_T = \chi,$$

(3.73)

while for ferromagnets

$$\left(\frac{\partial j}{\partial H}\right)_T = \chi + T\left(\frac{\partial\chi}{\partial H}\right)_T,$$

(3.74)

i.e., for ferromagnets it is necessary to take account of the dependence of χ on H in calculating $(\partial j/\partial H)_T$.

Consequently for paramagnets and ferromagnets

$$\left(\frac{\partial j}{\partial T}\right)_H < 0 \quad \text{and} \quad \left(\frac{\partial H}{\partial T}\right)_j > 0.$$

(3.75)

Using these relations we find from (3.62) that for paramagnets and ferromagnets we always have

$$c_{H,\,p} > c_{j,\,p}.$$

(3.76)

For ideal paramagnets $(C_{H,\,p} - C_{j,\,p})$ can be calculated explicitly. Using the equation for the Curie law (3.13), we find

$$\left(\frac{\partial\chi}{\partial T}\right)_H = \left(\frac{\partial\chi}{\partial T}\right)_j = -\frac{A}{T^2} \quad \text{and} \quad \left(\frac{\partial\chi}{\partial H}\right)_T = 0.$$

(3.77)

Then from (3.64)-(3.66) we find

$$\left(\frac{\partial j}{\partial T}\right)_H = -\frac{AH}{T^2},$$

(3.78)

$$\left(\frac{\partial j}{\partial H}\right)_T = \frac{A}{T},$$

(3.79)

and

$$\left(\frac{\partial H}{\partial T}\right)_j = \frac{H}{T}.$$

(3.40)

Substituting the partial derivatives into Eqs. (3.60), (3.62), and (3.63), we find for an ideal paramagnet

$$c_{H,p} - c_{j,p} = \frac{AH^2}{T^2} \tag{3.80}$$

or, using (3.10) and (3.13),

$$c_{H,p} - c_{j,p} = \frac{Hj}{T}. \tag{3.81}$$

We can draw a definite analogy between the relation which connects $c_{H,p}$ and $c_{j,p}$ for an ideal paramagnet and the well-known relation which connects c_p and c_v for an ideal gas

$$c_p - c_v = R, \tag{3.82}$$

which with the aid of the Clapeyron equation can be represented in the form

$$c_p - c_v = \frac{pv}{T}. \tag{3.83}$$

It is not hard to see that the right-hand sides of Eqs. (3.82) and (3.83) have the same structure - the derivative of the generalized force with respect to the generalized coordinate divided by the temperature.

The relations between $c_{H,p}$ and H and $c_{j,p}$ and j can be obtained as follows.

From Eq. (3.56)

$$c_{H,p} = T \left(\frac{\partial s}{\partial T} \right)_{H,p};$$

evidently

$$\left(\frac{\partial c_{H,p}}{\partial H} \right)_{T,p} = T \left[\frac{\partial}{\partial H} \left(\frac{\partial s}{\partial T} \right)_{H,p} \right]_{T,p}. \tag{3.84}$$

Since in calculating mixed derivatives the result of the differentiation does not depend on the order of differentiating, Eq. (3.84) can be written in the form

$$\left(\frac{\partial c_{H,p}}{\partial H} \right)_{T,p} = T \left[\frac{\partial}{\partial T} \left(\frac{\partial s}{\partial H} \right)_{T,p} \right]_{H,p}. \tag{3.85}$$

Using (3.31) it follows that

$$\left(\frac{\partial c_{H,p}}{\partial H} \right)_{T,p} = T \left(\frac{\partial^2 j}{\partial T^2} \right)_{H,p}, \tag{3.86}$$

which relates the change in the heat capacity $c_{H,\,p}$ along an isotherm with changing magnetic field strength to the derivative $(\partial^2 j/\partial T^2)_{H,\,p}$.

Integrating (3.86), we find

$$c_{H,p}(H,T,p) - c_{H,p}(H=0,T,p) = T \int_0^H \left(\frac{\partial^2 j}{\partial T^2}\right)_{H,p} dH, \qquad (3.87)$$

where the integral which enters this equation is calculated along an isotherm T = const (with p = const).

Since

$$c_{H,p}(H=0,T,p) = c_p(T,p), \qquad (3.88)$$

Eq. (3.87) can be written in the form

$$c_{H,p}(H,T,p) - c_p(T,p) = T \int_0^H \left(\frac{\partial^2 j}{\partial T^2}\right)_{H,p} dH. \qquad (3.89)$$

This equation shows that the isobaric heat capacity of a magnet changes with magnetic field.

Using (3.10) this equation can be transformed to

$$c_{H,p}(H,T,p) - c_p(T,p) = T \int_0^H \left(\frac{\partial^2 \chi}{\partial T^2}\right)_{H,p} H dH. \qquad (3.90)$$

It follows from this equation that for diamagnets, in which in agreement with (3.67) we have $(\partial\chi/\partial T)_{H,\,p} = 0$,

$$c_{H,\,p} = c_p, \qquad (3.91)$$

i.e., the heat capacity of a diamagnet does not change with a change of the magnetic field strength. It now becomes possible to understand the previous conclusion that the heat capacities $c_{H,p}$ and $c_{j,p}$ are the same in diamagnets [Eq. (3.69)].

For ideal paramagnets we find from (3.77)

$$\left(\frac{\partial\chi}{\partial T}\right)_{H,\,p} = -\frac{\Lambda}{T^2}$$

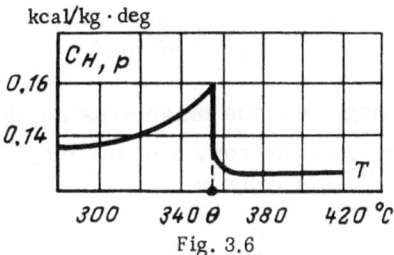

Fig. 3.6

so that

$$\left(\frac{\partial^2 \chi}{\partial T^2}\right)_{H,p} = \frac{2A}{T^3}.$$

(3.92)

Substituting this value into Eq. (3.90), we find

$$c_{H,p}(H, T, p) - c_p(T, p) = \frac{AH^2}{T^2}$$

(3.93)

or, using (3.13),

$$c_{H,p} - c_p = \frac{\chi H^2}{T}.$$

(3.94)

Using (3.10) we write this equation as follows:

$$c_{H,p} - c_p = \frac{Hj}{T}.$$

(3.95)

As we see from a comparison of this relation with (3.81), $c_{j,p}$ and c_p differ for ideal paramagnets.

In speaking of the heat capacity of a magnet is should be mentioned that in ferromagnets the heat capacity $c_{H,p}$, like a number of other thermodynamic quantities, passes through a maximum at the Curie point (see Fig. 3.6, in which we show the heat capacity of nickel). The temperature dependence of the heat capacity of an antiferromagnet behaves similarly during passage through the antiferromagnetic Curie point (in Fig. 3.7 we show the temperature dependence of $c_{H,p}$ for anhydrous $MnCl_2$).

3.4. Thermodynamic Processes in Magnets

We will consider the properties of the basic reversible processes in magnets.

Isothermal Processes. If the magnetic field strength H (and consequently the magnetization j) varies under isothermal conditions with the temperature of the magnet held constant (for example, the magnet may be placed in a heat bath), then, as we see from the Maxwell equation (3.31),

$$\left(\frac{\partial s}{\partial H}\right)_{T,p} = \left(\frac{\partial j}{\partial T}\right)_{H,p},$$

and if $(\partial j/\partial T)_{H,\,p}$ is negative (the magnetization decreases with increasing temperature), the entropy s of the magnet decreases with increasing magnetic field strength:

$$\left(\frac{\partial s}{\partial H}\right)_{T,p} < 0. \tag{3.96}$$

This rule is valid for paramagnets and ferromagnets. Since $(\partial j/\partial T)_H = 0$ for diamagnets in agreement with (3.68) the entropy of a diamagnet does not change with magnetic field strength.

The change in the entropy of a magnet in an isothermal process (and with p = const) can be calculated using the relation

$$s_2(H_2,T) - s_1(H_1,T) = \int_{H_1}^{H_2} \left(\frac{\partial s}{\partial H}\right)_{T,p} dH. \tag{3.97}$$

Using Eq. (3.31) we can put this equation in the form

$$s_2(H_2,T) - s_1^{'}(H_1,T) = \int_{H_1}^{H_2} \left(\frac{\partial j}{\partial T}\right)_{H,p} dH. \tag{3.98}$$

In view of (3.75) it is obvious that the entropy decreases along an isotherm with increasing H for paramagnets and ferromagnets.

Finally, using (3.10), this equation can be transformed to

$$s_2(H_2,T) - s_1(H_1,T) = \int_{H_1}^{H_2} \left(\frac{\partial \chi}{\partial T}\right)_{H,p} H dH. \tag{3.99}$$

For the particular case of an ideal paramagnet we can use this equation and (3.77) to find

$$s_2(H_2, T) - s_1(H_1, T) = -\frac{A(H_2^2 - H_1^2)}{2T^2}. \tag{3.100}$$

The amount of heat q_{1-2} going into a unit weight of material (or going out from it) in an isothermal process, as we know can be calculated as

$$q_{1-2} = T(s_2 - s_1), \tag{3.101}$$

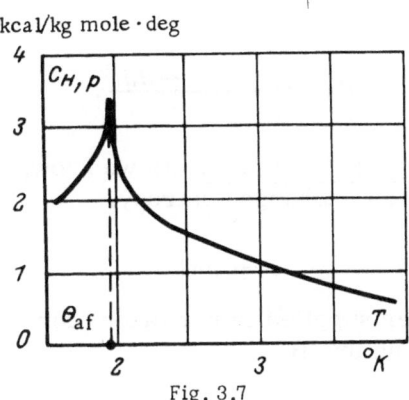

Fig. 3.7

where $(s_2 - s_1)$ is the change in the entropy of the body in an isothermal process. As shown above, the entropy of a paramagnet or a ferromagnet decreases with increasing magnetic field strength. In view of this it follows from (3.101) that heat is removed from the magnet with increasing H ($s_2 < s_1$ and $q_{1-2} < 0$); on the other hand, the magnet will absorb heat ($q_{1-2} > 0$) with a decrease in the strength of the external. For a diamagnet $q_{1-2} = 0$ in an isothermal process in agreement with this.

The work done by the magnetic field on a magnet in an isothermal process is given by

$$l^*_{1-2} = -\int_{j_1}^{j_2} H \, dj \tag{3.102}$$

as follows from (3.18), or, using (3.10), by

$$l^*_{1-2} = -\int_{j_1}^{j_2} \frac{j}{\chi} \, dj. \tag{3.103}$$

Here the integrals are calculated along an isotherm, T = const.

For those kinds of magnet in which the magnetic susceptibility χ does not change with H or j, i.e., for diamagnets and paramagnets, we find

$$l^*_{1-2} = -\frac{j_2^2 - j_1^2}{2\chi} \tag{3.104}$$

or, what is the same,

$$l^*{}_{1-2} = -\frac{H_2 j_2 - H_1 j_1}{2}. \qquad (3.105)$$

Adiabatic Processes. As we know, a reversible adiabatic process has unchanged entropy.

$$s = \text{const.} \qquad (3.106)$$

Since heat is neither supplied to nor taken from the system in an adiabatic process, naturally

$$q_{1-2} = 0. \qquad (3.107)$$

It is to be understood that in an adiabatic process the temperature of the magnet varies with changing magnetic field strength. It is clear that an increase of H will be accompanied by an increase in the temperature of the magnet; the heat liberated will go into heating the magnet itself; a decrease of H under adiabatic conditions will naturally decrease the temperature of the magnet.

Questions involving changes in the temperature of a magnet in an adiabatic process will be discussed in detail in a subsequent section.

The work done by a magnetic field on a magnet in an adiabatic process is given by equations similar to (3.102) and (3.103) above, with the difference that in this case the integrals which enter these equations should be calculated not along isotherms but along isentropes.

Processes with H = const (with p = const). In a process with H = const and p = const the magnetization j of the magnet varies with temperature (if the magnetic susceptibility of the magnet depends on the temperature). The amount of heat supplied to the magnet (or given up by it) in this process is determined from Eq. (3.25):

$$T ds = di^* - v dp + j dH.$$

Since

$$dq = T ds, \qquad (3.108)$$

for the case considered (dH = 0, dp = 0) we find from (3.25)

$$q_{1-2} = i^*_2 - i^*_1. \tag{3.109}$$

The changes in the entropy of a magnet in the given process can be calculated using the obvious relation

$$s_2(H,T_2) - s_1(H,T_1) = \int_{T_1}^{T_2} \left(\frac{\partial s}{\partial T}\right)_{H,p} dT, \tag{3.110}$$

which, using (3.56), is written in the form

$$s_2(H,T_2) - s_1(H,T_1) = \int_{T_1}^{T_2} \frac{c_{H,p}}{T} dT. \tag{3.111}$$

Evidently the entropy of a magnet always increases with temperature in a process with H = const (with p = const).

The work done by the magnetic field on the magnet in this process is determined using the general relation (3.102) as

$$l^*_{1-2} = - \int_{j_1}^{j_2} H dj,$$

from which we find for the given process

$$l^*_{1-2} = -(Hj_2 - Hj_1) \tag{3.112}$$

or, what is the same,

$$l^*_{1-2} = -H^2 [\chi(H,T_2) - \chi(H,T_1)], \tag{3.113}$$

which is proportional to the difference between the magnetic susceptibilities in the initial and final states of the process.

It is thus evident that for diamagnets, in which the magnetic susceptibility does not vary with temperature, $l^*_{1-2} = 0$.

For paramagnets and ferromagnets, it is clear using (3.70) that if $T_2 > T_1$ then $\chi(H, T_2) < \chi(H, T_1)$ and consequently $l^*_{1-2} > 0$, i.e., work is done by the magnet.

For the particular case of an ideal paramagnet, we can use (3.13) to find from (3.113)

$$l^*_{1-2} = AH^2 \left(\frac{1}{T_1} - \frac{1}{T_2} \right). \tag{3.114}$$

Process with j = const (with p = const). In such a process the changes in the temperature of the magnet and the strength of the magnetic field should occur in such a way that the magnetization j remains constant as these two quantities change simultaneously.

The amount of heat supplied to (or given up by) the magnet in processes with j = const and p = const is given by Eq. (3.22) as

$$T ds = du + p dv - H dj,$$

which can be transformed using

$$p dv = d(pv) - v dp$$

into the form

$$T ds = d(u + pv) - v dp - H dj. \tag{3.115}$$

Since dj = 0 and dp = 0 in the given case, we find [using (3.108)] that

$$q_{1-2} = i_2 - i_1 \tag{3.116}$$

is an amount of heat equal to the difference in the "ordinary" enthalpies at the end and beginning of the process.†

The change of entropy in the process of interest is calculated using the obvious relation

$$s_2(j, T_2) - s_1(j, T_1) = \int_{T_1}^{T_2} \left(\frac{\partial s}{\partial T} \right)_{j,p} dT, \tag{3.117}$$

---

†It should be noted that in the presence of a magnetic field this "ordinary" enthalpy of a magnet i = u + pv differs from the "ordinary" enthalpy in the absence of a field since, as noted previously, when the magnetic field strength changes the internal energy u of the magnet also changes.

which can be written in the form

$$s_2(j,T_2) - s_1(j,T_1) = \int_{T_1}^{T_2} \frac{c_{j,p}}{T} \, dT \qquad (3.118)$$

using Eq. (3.56).

As for the work done by the magnetic field on the magnet, we see from the general relation (3.102) that in the process of interest this work is identically zero:

$$l^*_{1-2} = 0. \qquad (3.119)$$

3.5. The Magnetocaloric, Magnetostrictive, and Magnetoelastic Effects

By the magnetocaloric effect we mean the change in the temperature of a magnet with a change in the strength of the external magnetic field.

From what has been said in the preceding section it follows that the magnetocaloric effect will occur in any thermodynamic process (where the field strength at the magnet changes) with the exception of isothermal processes. Among these processes we are particularly interested in the magnetocaloric effect with an adiabatic change in the state of the magnet.

The change in the temperature of a magnet brought about by the magnetocaloric effect in an adiabatic process is determined from the relation

$$T_2 - T_1 = \int_{H_1}^{H_2} \left(\frac{\partial T}{\partial H} \right)_{s,p} dH. \qquad (3.120)$$

The partial derivative $(\partial T/\partial H)_{s,p}$ which enters this equation is calculated as follows.

In agreement with (1.48) we can write

$$\left(\frac{\partial T}{\partial H} \right)_{s,p} = - \left(\frac{\partial s}{\partial H} \right)_{T,p} \left(\frac{\partial T}{\partial s} \right)_{H,p}. \qquad (3.121)$$

This relation can be put in the form

$$\left(\frac{\partial T}{\partial H}\right)_{s,p} = -\frac{T}{c_{H,p}}\left(\frac{\partial j}{\partial T}\right)_{H,p}. \tag{3.122}$$

using (3.31) and (3.56).

Since the heat capacity $c_{H,p}$ is always positive, the sign of the derivative $(\partial T/\partial H)_{s,p}$ is determined by the sign of $(\partial j/\partial T)_{H,p}$.

For diamagnets, in agreement with (3.68) we find from (3.122) that

$$\left(\frac{\partial T}{\partial H}\right)_{s,p} = 0 \tag{3.123}$$

When the magnetic field strength changes adiabatically, the temperature of a diamagnet remains the same.

For paramagnets and ferromagnets, in agreement with (3.75) we find

$$\left(\frac{\partial j}{\partial T}\right)_{H,p} < 0,$$

i.e., the magnetization decreases with increasing temperature. Consequently in this case as we see from (3.122) that

$$\left(\frac{\partial T}{\partial H}\right)_{s,p} > 0. \tag{3.124}$$

This means that if we adiabatically isolate a magnet, its temperature increases with increasing strength of the external magnetic field in which it is located (i.e., with the magnetization of the magnet); correspondingly with decreasing magnetic field strength (demagnetization) the temperature decreases.

Using (3.122) we can put Eq. (3.120) into the form

$$T_2 - T_1 = -\int_{H_1}^{H_2} \frac{T}{c_{H,p}}\left(\frac{\partial j}{\partial T}\right)_{H,p} dH \tag{3.125}$$

or, using (3.10),

$$T_2 - T_1 = -\int_{H_1}^{H_2} \frac{TH}{c_{H,p}}\left(\frac{\partial \chi}{\partial T}\right)_{H,p} dH. \tag{3.126}$$

The requirements for adiabatic magnetization (or demag-
natization) can be r e a l i z e d to a high degree of approximation
experimentally.

The temperature change $\Delta T = T_2 - T_1$ with an adiabatic change
in the magnetic field strength H turns out to be quite substantial
for a number of ferromagnets and paramagnets. As an example,
Fig. 3.8 shows experimental data on the magnetocaloric effect in
nickel. The graph shows the dependence of ΔT on H (the field
strength is varied from H = 0) for various initial temperatures.
As we see from this graph, with fields of 15 to 20 kOe, ΔT can
exceed 1°C in nickel.

The magnetocaloric effect in some paramagnetic materials
is used to produce superlow temperatures using the so-called
adiabatic demagnetization method, which is described in detail in
the next section.

In addition to the magnetocaloric effect, the m a g n e t o-
s t r i c t i v e e f f e c t is observed when a body is magnetized. By
magnetostriction we mean the change in the size of a magnet with
a change in the external magnetic field strength.

As shown by experiment, the size of a magnet can increase
or decrease with increasing H; the magnitude and sign of the mag-

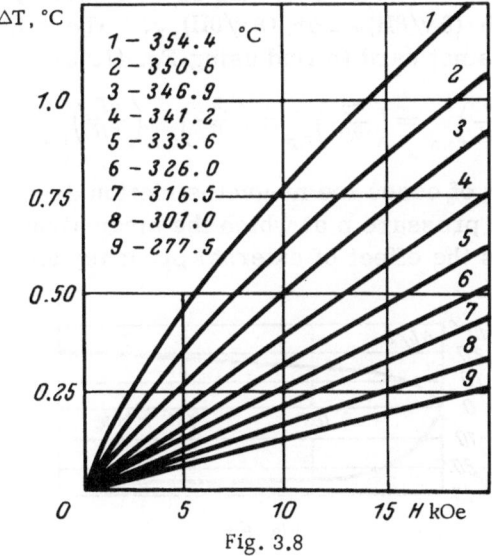

Fig. 3.8

netostrictive effect are individual characteristics of the given
magnet. Figure 3.9 shows data on the magnetostriction in an iron–
nickel–cobalt alloy (invar) and in pure nickel; in this graph the
ordinate gives the magnitude of the relative change in sample size
$\Delta l / l$, where l is the sample size at H = 0 and Δl is the change in
the sample length when the magnetic field strength changes from
H = 0 to the given value. As we see from this graph, the size of
the alloy sample (curve 2) increases with increasing H (the sample
expands) while the dimensions of the pure nickel sample (curve 1)
contract with increasing H. Here $\Delta l / l$ itself turns out to be quite
appreciable.

For a number of ferromagnets we observe anisotropy of the
magnetostriction with fields H < H_s (see Fig. 3.1) — the magnitude
(and in some cases even the sign) of the magnetostriction is
different for different crystallographic axes. In ferromagnets with
H > H_s and in most paramagnets the magnetostriction is isotropic —
the changes in the dimensions of the magnet are the same in all
directions.

The magnetostriction effect at constant external pressure p
is characterized by the derivative $\partial v / \partial H$ (while for anisotropic
magnetostriction it is characterized by the derivative $\partial l / \partial H$).
Here we should distinguish the conditions, isothermal or adiabatic,
of the magnetostriction and, depending on this choice, operate,
respectively, with $(\partial v / \partial H)_{T,p}$ or $(\partial v / \partial H)_{s,p}$. The relation between
these quantities is not hard to find using Eq. (1.49):

$$\left(\frac{\partial v}{\partial H}\right)_{s,p} = \left(\frac{\partial v}{\partial H}\right)_{T,p} + \left(\frac{\partial v}{\partial T}\right)_{H,p}\left(\frac{\partial T}{\partial H}\right)_{s,p} \tag{3.127}$$

In a number of cases the magnetostriction depends substantially on the external pressure p at which the magnetization is done.
Figure 3.10 shows the effect of external pressure (compression of

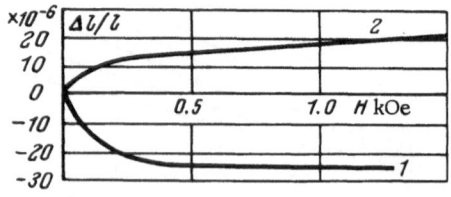

Fig. 3.9

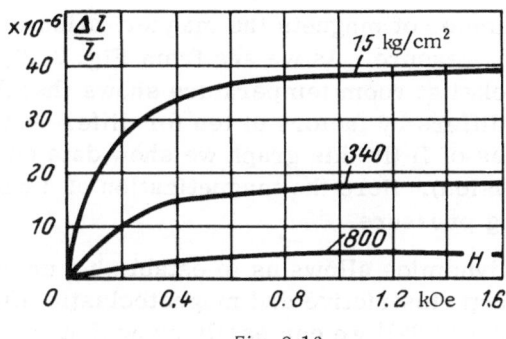

Fig. 3.10

the sample) on the magnetostriction of nickel. As we see from this graph, the relation $\Delta l/l = f(H)$ turns out to be substantially different for different isobars.

The magnetostriction also depends substantially on the temperature of the magnet (Fig. 3.11).

The magnetoelastic effect is an important effect seen in magnets. This is the change in the magnetization with a change in the external pressure. The magnetoelastic effect at constant external magnetic field strength is characterized by the derivative $\partial j/\partial p$. Just as for magnetostriction, in the magnetoelastic effect we should distinguish whether isothermal or adiabatic conditions prevail. Depending on which conditions occur, we should use $(\partial j/\partial p)_{T,H}$ or $(\partial j/\partial p)_{s,H}$ to characterize the magnetoelastic effect. We can use Eq. (1.49) to find the relation between these quantities:

$$\left(\frac{\partial j}{\partial p}\right)_{s,H} = \left(\frac{\partial j}{\partial p}\right)_{T,H} + \left(\frac{\partial j}{\partial T}\right)_{p,H}\left(\frac{\partial T}{\partial p}\right)_{s,H}. \tag{3.128}$$

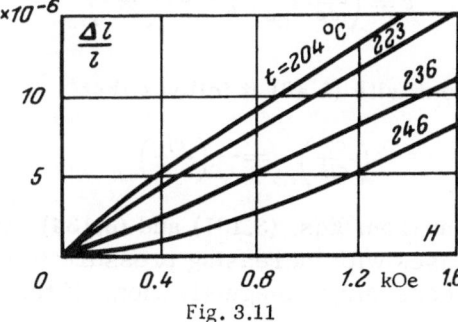

Fig. 3.11

For a number of magnets the magnetization is a quite strong function of the pressure. As we see from Fig. 3.12, the relation $j = f(H)$ for nickel at room temperature shows that the magnetization of nickel differs by factors of ten for different isobars for the same values of H (in this graph we show data obtained for samples in tension). Here the magnetization of nickel decreases with increasing pressure.

Thermodynamics allows us to establish a unique relation between the magnetostrictive and magnetoelastic effects. From Eq. (3.26) [using (3.22)] we can easily show that

$$d\varphi^* = -\, sdT + vdp - jdH. \tag{3.129}$$

Since φ^* is a state function, $d\varphi^*$ is a total differential and consequently

$$v = \left(\frac{\partial \varphi^*}{\partial p}\right)_{T,H}, \quad j = -\left(\frac{\partial \varphi^*}{\partial H}\right)_{T,p}. \tag{3.130}$$

In agreement with Eq. (1.43) we find that

$$\left(\frac{\partial v}{\partial H}\right)_{T,p} = -\left(\frac{\partial j}{\partial p}\right)_{T,H}. \tag{3.131}$$

Similarly we find from Eq. (3.23), using (3.22),

$$di^* = Tds + vdp - jdH, \tag{3.132}$$

from which it is evident that

$$v = \left(\frac{\partial i^*}{\partial p}\right)_{s,H}, \quad j = -\left(\frac{\partial i^*}{\partial H}\right)_{s,p}. \tag{3.133}$$

In agreement with (1.43) it follows that

$$\left(\frac{\partial v}{\partial H}\right)_{s,p} = -\left(\frac{\partial j}{\partial p}\right)_{s,H}. \tag{3.134}$$

It is evident from Eqs. (3.131) and (3.134) that the magnetization decreases with increasing pressure in magnets whose dimensions increase during magnetization. If the dimensions of the magnet contract during magnetization then j increases with increasing pressure.

The relation for the change in the volume of a magnet due to magnetostriction is written as follows:

$$v(T, p, H) - v(T, p, H = 0) = \int_0^H \left(\frac{\partial v}{\partial H}\right)_{T, p} dH \qquad (3.135)$$

and

$$v(s, p, H) - v(s, p, H = 0) = \int_0^H \left(\frac{\partial v}{\partial H}\right)_{s, p} dH. \qquad (3.136)$$

Using (3.131) and (3.134) we find correspondingly

$$v(T, p, H) - v(T, p, H = 0) = -\int_0^H \left(\frac{\partial j}{\partial p}\right)_{T, H} dH \qquad (3.137)$$

and

$$v(s, p, H) - v(s, p, H = 0) = -\int_0^H \left(\frac{\partial j}{\partial p}\right)_{s, H} dH. \qquad (3.138)$$

Since it follows from (3.10) that

$$\left(\frac{\partial j}{\partial p}\right)_H = H \left(\frac{\partial \chi}{\partial p}\right)_H, \qquad (3.139)$$

while $(\partial\chi/\partial p)_H$ can be regarded as practically independent of H for most cases,

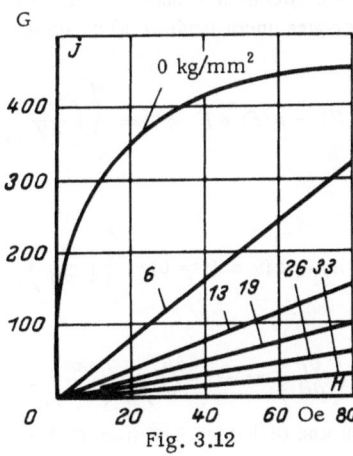

Fig. 3.12

$$v(T, p, H) - v(T, p, H = 0) = -\frac{H^2}{2}\left(\frac{\partial \chi}{\partial p}\right)_{T, H} \tag{3.140}$$

and

$$v(s, p, H) - v(s, p, H = 0) = -\frac{H^2}{2}\left(\frac{\partial \chi}{\partial p}\right)_{s, H}. \tag{3.141}$$

It is further evident that a change in the volume of the magnet due to an external magnetic field can be formally described as a change in v due to a certain additional pressure Δp, i.e.,

$$v(p, H) - v(p, H = 0) = v(p + \Delta p, H = 0) - v(p, H = 0). \tag{3.142}$$

Clearly,

$$v(p + \Delta p) - v(p) = \int_{p}^{p+\Delta p} \left(\frac{\partial v}{\partial p}\right)_T dp \approx \left(\frac{\partial v}{\partial p}\right)_T \Delta p. \tag{3.143}$$

From a comparison of (3.143), (3.140), and (3.141) we see that this additional external "magnetic pressure" is given by the following relations:

$$\Delta p_T = -\frac{H^2}{2}\left(\frac{\partial \chi}{\partial v}\right)_{T, H} \tag{3.144}$$

and

$$\Delta p_s = -\frac{H^2}{2}\left(\frac{\partial \chi}{\partial v}\right)_{s, H}. \tag{3.145}$$

If the volume of the magnet is held constant with a change in the external magnetic field strength (v = const) then a change in H naturally changes the pressure in the magnet. When H changes under isothermal or adiabatic conditions we have, respectively,

$$p(T, v, H) - p(T, v, H = 0) = \int_{0}^{H} \left(\frac{\partial p}{\partial H}\right)_{T, v} dH \tag{3.146}$$

and

$$p(s, v, H) - p(s, v, H = 0) = \int_{0}^{H} \left(\frac{\partial p}{\partial H}\right)_{s, v} dH. \tag{3.147}$$

Since

$$\left(\frac{\partial p}{\partial H}\right)_{v} = -\left(\frac{\partial p}{\partial v}\right)_{H}\left(\frac{\partial v}{\partial H}\right)_{p}, \tag{3.148}$$

Using (3.131), (3.134), and (3.10) we find from (3.146) and (3.147) that

$$p(T, v, H) - p(T, v, H = 0) = \int_0^H \left(\frac{\partial \chi}{\partial v}\right)_{T, H} H dH \qquad (3.149)$$

and

$$p(s, v, H) - p(s, v, H = 0) = \int_0^H \left(\frac{\partial \chi}{\partial v}\right)_{s, H} H dH. \qquad (3.150)$$

Since $(\partial \chi / \partial v)_H$ usually varies weakly with a change in H, we have

$$p(T, v, H) - p(T, v, H = 0) = \frac{H^2}{2}\left(\frac{\partial \chi}{\partial v}\right)_{T, H} \qquad (3.151)$$

and

$$p(s, v, H) - p(s, v, H = 0) = \frac{H^2}{2}\left(\frac{\partial \chi}{\partial v}\right)_{s, H}. \qquad (3.152)$$

These relations differ in sign from Eqs. (3.144) and (3.145). This is not surprising, since Eqs. (3.144) and (3.145) contain an additional pressure of the external medium (fictitious) while Eqs. (3.151) and (3.152) contain the change of pressure in the magnet with v = const. These quantities have different signs: for example, if an increase of H causes the volume of a gaseous paramagnet to increase, then this effect can be formally interpreted as an expansion of the magnet resulting from a decrease in the external pressure [Eqs. (3.144) and (3.145)]. If, however, the volume of the magnet is held constant, then the pressure in the magnet will grow [Eqs. (3.151) and (3.152)].

The pressure change in the magnet with a change in the external magnetic field strength can reach substantial values in a number of cases. This effect is used, in particular, for forming ferromagnetic materials using strong magnetic fields.

3.6. Adiabatic Demagnetization

As is well known, the cooling agent having the lowest boiling point is liquid helium; its boiling point at atmospheric pressure is 4.20°K. By intense pumping of the liquid helium vapors, i.e., by substantially lowering the pumped helium's vapor pressure, one can obtain a temperature of $\sim 0.71°K.$[†]

How can still lower temperatures be produced? In 1926 P. Debye and W. Giauque proposed to use the magnetocaloric effect for this purpose.

As shown in the preceding section, under adiabatic conditions (s = const) the temperature of a paramagnet increases if we increase the strength of the external magnetic field in which it is located; correspondingly the temperature will decrease [Eq. (3.124)]

[†]The saturated vapor pressure of helium at this temperature is 3.6×10^{-3} torr.

with a decrease in the magnetic field strength (demagnetization).

We note that for those paramagnets for which the Curie law (3.13) is valid, we find, in agreement with Eq. (3.78),

$$\left(\frac{\partial j}{\partial T}\right)_{H,p} = -\frac{AH}{T^2} .$$

In view of this relation, Eq. (3.122)

$$\left(\frac{\partial T}{\partial H}\right)_{s,p} = -\frac{T}{c_{H,p}}\left(\frac{\partial j}{\partial T}\right)_{H,p}$$

assumes the form

$$\left(\frac{\partial T}{\partial H}\right)_{s,p} = \frac{AH}{c_{H,p}T} . \qquad (3.153)$$

For ordinary temperatures, where $(\partial j/\partial T)_{H,p}$ is comparatively small in agreement with (3.78) (since T^2 is large) while the heat capacity $c_{H,p}$ on the other hand is substantial, $(\partial T/\partial H)_{s,p}$ turns out to be quite small. At low temperatures the Curie law is usually inapplicable. However, in 1923 it was found that a relation such as (3.13) extends to liquid helium temperatures for some paramagnetic salts. It was just this fact which allowed Debye and Giauque to propose the use of adiabatic demagnetization of certain paramagnetic salts to produce temperatures below 1°K.

As is well known, the heat capacities of materials decrease rapidly† at low temperatures, going to zero at absolute zero (in agreement with the Nernst theorem). Thus for low temperatures the denominator on the right-hand side of Eq. (3.153) increases sharply.

If T_1 is the initial temperature of a sample of a paramagnetic salt in a magnetic field of strength H, then the temperature T_2 which the sample acquires after an adiabatic decrease of the field strength to zero is determined as follows from Eq. (3.120) for the magnetocaloric effect:

$$T_2 = T_1 - \int_0^H \left(\frac{\partial T}{\partial H}\right)_{s,p} dH. \qquad (3.154)$$

---

†According to Debye's law, the heat capacity of solids at low temperatures varies as T^3.

Substituting $(\partial T/\partial H)_{s,\,p}$ from Eq. (3.153), we find (due to the small difference between T_1 and T_2)

$$T_2 = T_1 - \frac{AH^2}{2c_{H,p}T\text{av}},$$

(3.155)

where $T_{av} = (T_1 + T_2)/2$ and $c_{H,p}$ is the average heat capacity in this temperature range.

Equation (3.155) allows us to estimate the temperature which will be reached during adiabatic demagnetization of a paramagnetic salt, from known values of T_1, H, $c_{H,p}$, and A.

Since near absolute zero the heat capacity of a solid is proportional to the cube of the temperature, in agreement with Debye's law,

$$c_{H,\,p} = bT^3,$$

(3.156)

where b is a constant, it follows from Eq. (3.155) that

$$T_1 - T_2 \sim \frac{1}{T^4},$$

(3.157)

i.e., $(T_1 - T_2)$ attainable for the same value of H increases sharply with a decrease of the initial temperature T_1.

The first experiment on adiabatic demagnetization was done in 1933. In this experiment a temperature of 0.53°K was reached by demagnetizing a paramagnetic salt — gadolinium sulfate (with an initial salt temperature of 3.4°K and a magnetic field strength of 8000 Oe). Since that time methods have been substantially perfected. The lowest temperature reached by adiabatic demagnetization is about 0.001°K.

Adiabatic demagnetization is carried out as follows (Fig. 3.13). The sample 1 of a paramagnetic salt is placed in vessel 2 loaded into a dewar vessel containing liquid helium. The helium dewar vessel 3 is placed in the gap between the poles of an electromagnet 4. The temperature of the liquid helium in the dewar vessel is lowered to $T_1 = 0.9$-1.3°K by pumping strongly on it. Then an external magnetic field is applied to the sample (by switching on the current in the electromagnet windings) and the sample is magnetized. During magnetization heat is given up by the sample (the magnetocaloric effect). The outflow of heat from the sample during magnetization is isothermal. Vessel 2 is then thoroughly evacuated. When the current in the coils of the electromagnet is switched off, the external magnetic field drops, and the temperature of the adiabatically isolated sample decreases in accord with Eq. (3.155). In Fig. 3.14 we show the cooling process using adiabatic demagnetization in a T–H diagram; here line AB shows the isothermal magnetization of the sample while line BC shows adiabatic demagnetization of the sample.

If the initial temperature of magnetization (T_1) is about 1°K, then in order to obtain $T_2 < 0.01$°K one usually uses a magnetic field with a strength H not less than 20,000 to 25,000 Oe.

It should be noted that just as low a temperature can be achieved for smaller values of H if we use so-called two-stage demagnetization, the principle of which is as follows. We use two samples of the paramagnetic salt. The first of them (I) is adiabatically demagnetized by the usual method and cooled to a temperature T_2^I which is equal, for example, to 0.1°K. This sample is then used to absorb heat given up during magnetization of the second sample (II). Thus the magnetization temperature of the second sample T_1^{II}, will be close to the temperature T_2^I reached during demagnetization of the first sample. Then the second sample is demagnetized. From Eq. (3.155) we see that because the demagnetization of the second sample begins at a substantially lower temperature than that of the first sample, a quite low T_2 value can be attained without too high values of H. Thus it has been possible by using a two-stage demagnetization to reach temperatures of 0.001°K using a magnetic field of only 9000 Oe (the first sample was iron aluminum alum, which was cooled to

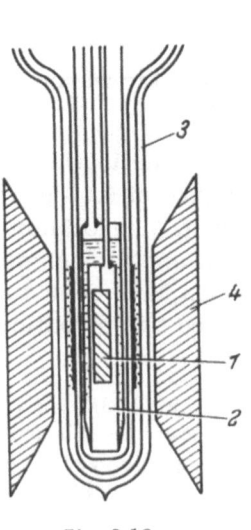

Fig. 3.13

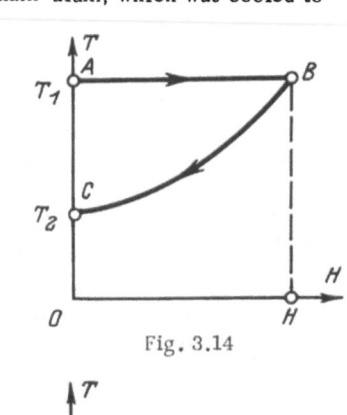

Fig. 3.14

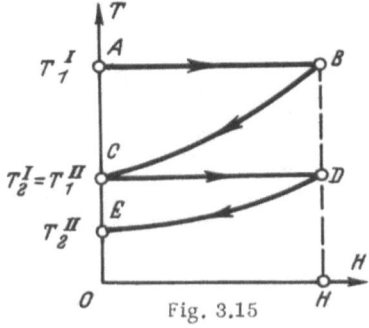

Fig. 3.15

0.25°K, and the second sample was a dilute chromium calcium alum). The cooling
process using two-step demagnetization is shown in a T–H diagram in Fig. 3.15; here
AB is an isothermal magnetization of the sample in the first step; BC is adiabatic
demagnetization of this sample; CD is isothermal magnetization of the sample for the
second step; and DE is adiabatic demagnetization of the sample for the second step.

It can thus be shown that adiabatic demagnetization can be used to produce
arbitrarily low temperatures (if, for example, we use still stronger magnetic fields or
use many demagnetization steps). However, this is not the case —adiabatic demagnetization does not allow one to produce temperatures below approximately 0.001°K.
The reason for this is as follows. The elementary magnets (magnetic dipoles) of a
paramagnet, which are randomly oriented in the absence of a field, lie along the field
when an external magnetic field is applied. Heat is given up (the magnetocaloric
effect). If we then magnetize the paramagnet (take away the external field), the
orientation of the magnetic dipoles again becomes disordered. This disordering of the
state of the magnetic dipoles occurs with absorption of heat, which is taken up from
the energy of the lattice vibrations of the paramagnet; because of this the temperature
of the paramagnet also drops.

For very low temperatures (of the order of 0.001°K) the magnetic dipoles begin
to interact among themselves; as a result of this interaction they lie parallel to each
other even in the absence of an external magnetic field — in agreement with the Nernst
theorem. The magnetocaloric effect vanishes here since the application of an external magnetic field does not cause heat to be liberated. The magnetic dipoles are
in fact already oriented in one direction.

This fact also determines the lower limit on the temperature which can be attained by adiabatic demagnetization.

An important independent problem is the establishment of
a temperature scale for temperatures below 1°K.

Ordinary methods are inapplicable in this temperature range.
Thus the temperature of liquid helium is most precisely determined by measuring the pressure p of the helium vapor above the
liquid helium. From the known p_s-T_s relation for helium we can
find T corresponding to a given value of p. This method becomes
useless at temperatures below 1°K; for helium p_s rapidly decreases
with decreasing temperature (at 0.5°K p_s is only about 10^{-5} torr;
precise measurement of such a low value of p_s is not practical).

It turns out that we can use Eq. (3.155) found above to calculate the temperature T_2 reached during adiabatic demagnetization. However, in reality this relation can only be used for crude
estimates of T_s, since T_s is very small and even a small error
in determining the initial temperature T_1 (prior to demagnetization)

or in the heat capacity $c_{H,p}$ of the salt can lead to large errors in calculating T_2 using Eq. (3.155).

Temperatures below 1°K are determined as follows. It is assumed that the Curie law (3.13) is valid down to the very lowest temperatures for the paramagnetic salt used in the demagnetization. By measuring the magnetic susceptibility χ_1 of the sample at a known temperature T_1 of the helium bath and the magnetic susceptibility χ_2 at an unknown temperature T* (the temperature which the sample reaches after demagnetization), we find, using Eq. (3.13),

$$T^* = \frac{\chi_1}{\chi_2}\, T_1.$$

(3.158)

The temperature T* calculated in this way is called the magnetic temperature.

As noted above, T* is calculated by extrapolating the Curie law down to the very lowest temperatures. Since the degree of validity of this extrapolation cannot be rigorously determined the question naturally arises how the magnetic temperature T* is related to the pure thermodynamic temperature T which enters into thermodynamic relations.[†] In other words, we must establish the form of the relation T = f(T*), bearing in mind that this relation will be different for different paramagnetic salts.

From the combined equation for the first and second laws of thermodynamics (for systems whose only form of work is work of expansion) written in the form

$$T\,ds = di - v\,dp,$$

it follows that

$$T = \left(\frac{\partial i}{\partial s}\right)_p.$$

(3.159)

This relation can be used as a definition of the thermodynamic temperature.

[†]It is to be understood that in the temperature range where the Curie law is rigorously valid, the thermodynamic temperature T agrees with the magnetic temperature T*.

Applied to the present case, we see that if i and s are, respectively, the enthalpy and entropy of a paramagnetic salt after demagnetization (i.e., with H = 0), then T is the true thermodynamic temperature of the salt after demagnetization.

It is also evident that Eq. (3.159) can be put in the form

$$T = \frac{\left(\frac{\partial i}{\partial T^*}\right)_p}{\left(\frac{\partial s}{\partial T^*}\right)_p}, \tag{3.160}$$

where T^* is the magnetic temperature. If we now find the relation between the enthalpy i and the entropy s for the salt and the quantity T^* at H = 0, we can use this relation, with the various given values of T^*, to calculate the temperature T corresponding to each T^* value.

We proceed as follows to calculate the derivative $(\partial i/\partial T^*)_p$ which enters Eq. (3.160).[†] Having lowered the temperature of a paramagnetic salt sample to a temperature T^* by demagnetization, we then supply heat to the sample from an external source until the sample temperature equals the temperature T_1 of the helium bath (i.e., the same temperature the sample had before magnetization). The heat is supplied with H = 0 under constant pressure conditions. The amount of heat needed to raise the sample temperature from T^* to T_1 can be precisely measured; clearly this amount of heat equals the enthalpy difference $q_{1-2} = i(T_1) - i(T^*)$. By measuring q for various values of T^* (with the same temperature T_1 of the helium bath) and then differentiating the relation $q = f(T^*)$ we can calculate $(\partial i/\partial T^*)_{p,H=0}$.

The derivative $(\partial s/\partial T^*)_p$ which enters the denominator on the right-hand side of Eq. (3.160) is calculated as follows.

It is clear that the difference in the entropy of the magnet in two states on an isotherm is given by a relation similar to (3.99):

[†] It should be stressed that the derivative $(\partial i/\partial T^*)_p$ is, strictly speaking, not the heat capacity C_p since the differentiation is done not with respect to the thermodynamic temperature T but with respect to the magnetic temperature T^* It is evident that

$$\left(\frac{\partial i}{\partial T^*}\right)_\nu = \left(\frac{\partial i}{\partial T}\right)_p \frac{dT}{dT^*} = c_p \frac{dT}{dT^*}.$$

$$s(T,H) = s(T,H=0) + \int_0^H \left(\frac{\partial \chi}{\partial T}\right)_{H,p} H dH \tag{3.161}$$

Here s(T, H) and s(T, H = 0) are the entropy of a magnet at a temperature T, respectively, in a field of strength H and with no field present.

Having the experimentally measured values of χ in the temperature range close to T_1 we can easily calculate the integral on the right-hand side of Eq. (3.161).

Knowing the temperature T_1 of the helium bath in which the paramagnetic salt sample is placed before demagnetization, we can use Eq. (3.161) to determine the entropy difference $[s(T_1,H) - s(T_1,H = 0)]$ for different external magnetic field strengths H. The isotherms T_1 = const in the coordinate system $[s(T_1,H) - s(T_1, H = 0)] = f(H)$ are shown schematically in Fig. 3.16.

Demagnetizing the sample to H = 0 several times, each time with a different initial field value H but for the same initial temperature T_1, we can use Eq. (3.158) to calculate T* for different values of H along this isotherm [as we see from (3.155), the greater the initial value of H for the same value of T, the lower the temperature produced by demagnetization]. Since the demagnetization process occurs adiabatically along an isentrope (Fig. 3.16), it is clear that the entropy of the sample at a temperature T_1 and in a field H (before demagnetization) equals the entropy at a temperature T* in the absence of a field (after demagnetization), i.e.,

$$s(T_1, H) = s(T^*, H=0). \tag{3.162}$$

Subtracting both sides of this equation from $s(T_1, H = 0)$, we find

$$s(T_1,H=0) - s(T_1,H) = s(T_1,H=0) - s(T^*,H=0), \tag{3.163}$$

whence, using (3.161),

$$s(T_1,H=0) - s(T^*,H=0) = -\int_0^H \left(\frac{\partial \chi}{\partial T}\right)_p H dH. \tag{3.164}$$

Since we know the temperature T* after demagnetization for each of the demagnetization processes which we have performed (i.e., for each initial value of H), we know the relation $[s(T_1, H = 0)$ $-s(T^*, H = 0)] = f(T^*)$.

Figure 3.17 shows such a relation for chromium potassium alum. Differentiating this relation we find the desired derivative $(\partial s/\partial T^*)_{p,\,H=0}$ [it should be borne in mind that $s(T_1, H = 0)$ does not change with a change of T^*, and thus its derivative with respect to T^* is zero].

Having obtained values for $(\partial i/\partial T^*)_{p,\,H=0}$ and $(\partial s/\partial T^*)_{p,\,H=0}$, we can use Eq. (3.160) to calculate the thermodynamic temperature T of the sample after demagnetization, corresponding to the magnetic temperature T^*.

It should be understood that T and T^* agree at the temperature T_1. The lower the temperature the greater the difference between T and T^*. For some paramagnetic salts at very low temperatures T and T^* differ by severalfold [see for example Fig. 3.18, in which we show the relation $T = f(T^*)$ for chromium potassium alum].

We will briefly treat the nature of the T−s diagram for a paramagnetic salt in the temperature range attained in adiabatic demagnetization. As an example, Fig. 3.19 shows the T−s diagram for iron ammonium alum constructed from experimental data. The diagram shows lines for H = const). As we see from this diagram the entropy of a paramagnetic salt depends strongly on the external magnetic field strength H. Here s decreases with increasing H along an isotherm, i.e., during magnetization; this

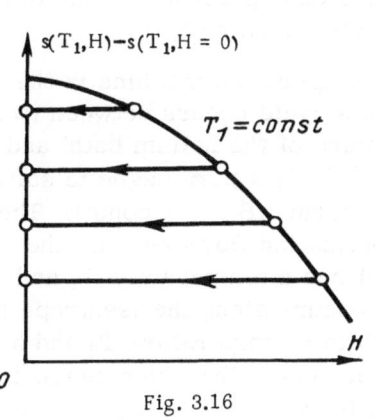

Fig. 3.16

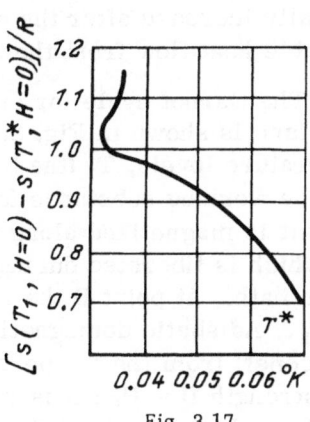

Fig. 3.17

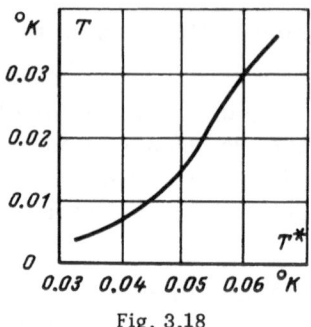

Fig. 3.18

regularity, which is common to all paramagnetic and ferromagnetic materials, was mentioned previously (p. 71). We also see from the diagram that the lines H = const bend back the other way on this diagram. This means that the heat capacity $c_{H,p}$ of iron ammonium alum given by Eq. (3.56),

$$c_{H,p} = T \left(\frac{\partial s}{\partial T} \right)_{H,p},$$

passes through a maximum at a temperature corresponding to the inflection point in the line H = const.

In conclusion we consider the scheme for the cycle of a refrigeration machine using adiabatic demagnetization. Such a refrigeration machine in continuous action allows one to maintain temperatures below 1°K for a prolonged period (while if we use a single demagnetization process the sample temperature will gradually increase after the end of the demagnetization, due to unavoidable heat flow from the surrounding medium).

The Carnot cycle for such a refrigeration machine in the T—s plane is shown in Fig. 3.20. This cycle occurs between two temperature levels, T_1 (the temperature of the helium bath) and T_2. The working substance for the cycle is a paramagnetic salt. The salt is magnetized along the isotherm AB (T_1 = const). The heat which is liberated during magnetization flows out into the helium bath. At point B the material has a temperature T_1 and $H = H_B$. Adiabatic demagnetization occurs along the isentrope BC (s_1 = const) from the temperature T_1 to a temperature T_2 and a field strength $H = H_C$; it is important to note that the demagnetization along the isentrope BC goes not to H = 0 but to $H_C > 0$. The

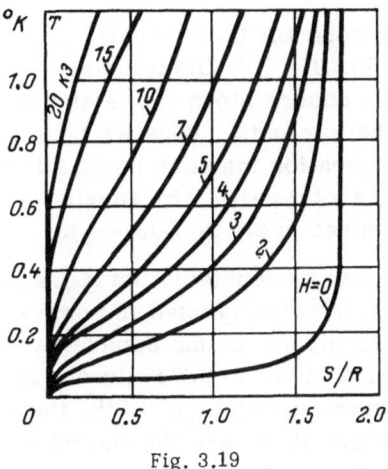

Fig. 3.19

isotherm CD (T_2 = const) corresponds to absorption of heat by the working substance from the cooled region; in this process the flow of heat into the salt is accompanied by a further drop in the strength of the external magnetic field (to H = 0) such that the increase in the temperature of the salt due to heat inflow is compensated by the temperature drop arising from demagnetization, and as a result the temperature T_2 remains constant along the line CD. After the working material reaches the state corresponding to point D on the diagram (T = T_2, H = 0) the working substance is adiabatically mag-

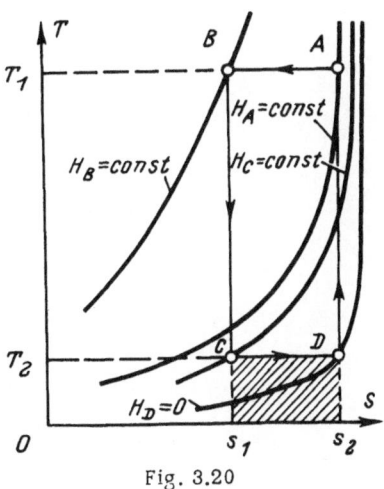

Fig. 3.20

netized along the isentrope DA (s_2 = const). After reaching the
state at point A (T = T_1, H = H_A) further magnetization (to H = H_B)
as already noted above occurs along the isotherm T_1 = const with
heat outflow into the helium bath. Thus we have a Carnot cycle
for a continuous refrigeration machine using adiabatic demagnetization of a paramagnetic salt. The amount of heat removed
from the cooled region per cycle is $T_2(s_1-s_2)$.

A cycle of this kind (undoubtedly deviating more or less from
the ideal Carnot cycle) is actually used in some experimental apparatus. Such continuously operating apparatus allows one to
maintain a cooled region at 0.2 to 0.3°K (in one such device 1.4×10^4
ergs of heat are removed per cycle at 0.3°K; the working material
is 15 g of iron ammonium alum, and the cooled region is a block of
chromium potassium alum of 15 g weight).

CHAPTER 4

Insulators

4.1. Introduction

All materials are either conductors or insulators with regard to their electrical properties. Insulators are materials having a low electrical conductivity. An external electric field does not affect the thermodynamic properties of a conductor, since it is well known that the field does not penetrate into it. However, the properties of an insulator change substantially when an electric field is applied, since the field penetrates into the insulator.

The processes which occur in an insulator when an electric field is applied can be simply illustrated as follows. An insulator consists of electrically neutral molecules. However, the molecules, which are electrically neutral as a whole, can have dipole moments.

There are two basic types of insulator: those whose molecules have a dipole moment in the absence of an external electric field (insulators with so-called polar molecules) and insulators whose molecules have zero dipole moment in the absence of a field but deform in such a way as to acquire a dipole moment when a field is applied (insulators with nonpolar molecules).

In the absence of an external electric field the dipole moments of polar molecules in an insulator are randomly oriented and their total electric moment is zero. When such an insulator is introduced into an electric field the dipole moments of the polar molecules are oriented along the field direction (the more "willingly," the greater the electric field strength E), and the total electric moment of the insulator is greater than zero.

For an insulator with nonpolar molecules, the dipole moments of the molecules are zero in the absence of an external electric field. When an external field is applied, these molecules, as noted above, deform and acquire a dipole moment (which increases with field strength E); these induced dipole moments are oriented along the field direction (the degree of orientation increases with increasing E), and the total electric moment of an insulator of this kind also becomes nonzero.

An insulator whose total electric moment is nonzero is described as polarized.

As a result of the polarization of the dipoles oriented along the field, an electric field of strength E' appears in the insulator, which combines with the external electric field of strength E_{ext} applied to the insulator. In other words, the resulting electric field strength in the insulator will be less than the strength of the external electric field in the absence of an insulator (E_{ext}): the polarized insulator gives an additional electric field of strength E' opposite to the initial field E_{ext}.

It should be stressed here that the additional field (of strength E') created by polarized charge can lead to a certain redistribution of the charges which create the external field E_{ext}. As a result, the strength of this external field varies; this in turn leads to a certain change in the field E', etc. In the final analysis (during an infinitestimal time interval) an equilibrium state is established in the system in which the strength of the resultant electric field E within the insulator is

$$E = D - E_{pol}, \tag{4.1}$$

where E_{pol} is the total electric field strength of the polarization charges and D is the strength of the external electric field in the absence of an insulator with the same charge distribution creating the field (i.e., charges which do not enter into the composition of the insulator) as is created in the presence of the insulator. We stress yet again that for the reasons mentioned above $D \neq E_{ext}$ and $E_{pol} \neq E'$.

The quantity D is called the electrostatic induction (or displacement).

The degree of polarization of the insulator is characterized by the magnitude of the polarization vector P, which is the electric moment per unit volume of the insulator,

$$P = \frac{\sum\limits_{i=1}^{n} p_i}{V}, \tag{4.2}$$

where $\sum\limits_{i=1}^{n} p_i$ is the geometric sum of the electric moments (p_i) of the molecules in an insulator occupying a volume V. Note that P is an extensive quantity.

As we know from physics, the magnitude of the polarization P in an insulator is related to the electric field E by

$$P = \alpha E, \tag{4.3}$$

where the coefficient α is called the electric polarization coefficient or the electric susceptibility. The value of α is an individual property of each insulating material. The electric polarization coefficient changes significantly with temperature; since an increase in the temperature allows disordering of the dipole orientation, α decreases with increasing temperature. As for the dependence of α on the electric field strength E, for most insulators the electric susceptibility is practically independent of E (for fields that are not too strong†).

The strength of the electric field E_{pol} of the polarization charges is uniquely related to the magnitude P of the polarization of the insulator. As we know from electrostatics, the relation between E_{pol} and P is as follows:

$$E_{pol} = 4\pi P. \tag{4.4}$$

---

†In strong electric fields the insulator saturates and further growth of E does not increase the polarization P. As we see from (4.3), this formally corresponds to a decrease of α with increasing E in this region of field strengths.

Substituting this value of E_{pol} into (4.1), we find

$$D = E + 4\pi P, \tag{4.5}$$

whence, using (4.4), we find

$$D = (1 + 4\pi\alpha)E \tag{4.6}$$

or

$$D = \varepsilon E. \tag{4.7}$$

The quantity

$$\varepsilon = 1 + 4\pi\alpha \tag{4.8}$$

is called the electric permeability of the insulator.

The electric susceptibility is always positive and consequently $\varepsilon > 1$ for any insulator. For vacuum, $\varepsilon = 1$.

We note in passing that if the electric susceptibility is small, i.e., ε is close to unity, the distortion of the external electric field E_{ext} due to the field of the polarization charges is quite small, so that we can take $E_{ext} \approx D$ to a good degree of precision.

As we see from what has been said above, there is a definite qualitative analogy between the characteristics of phenomena which occur in insulators and in magnets: the polarization in an isulator plays the same role as the magnetization for magnets — Eqs. (4.1) and (4.8) are analogous in their structure to Eqs. (3.1) to (3.8), with the electric susceptibility α corresponding to the magnetic susceptibility $\varkappa$ and the electric permeability ε corresponding to the magnetic permeability μ, etc.

We call attention to the fact that the quantity P in Eqs. (4.2) to (4.7) is the polarization per unit volume of the insulator, as follows from (4.2), and finally that the electric susceptibility α and the electric permeability ε also refer to a unit volume of the insulator.

It is clear that the total electric moment of the insulator can be written as

$$\mathfrak{P} = PV, \tag{4.9}$$

where V is the volume of the insulator.

Further, we can in principle introduce the idea of a specific (per unit weight) polarization and specific electric susceptibility as in Chapter 3 with regard to the specific magnetic susceptibility. However, it is the common practice to write the relations for insulators not in terms of weight-specific values for a specific electric susceptibility, but to use the electric permeability ε (and consequently volume-specific values). The reason for the use of this system is that ε is an important characteristic property of an insulator which in addition to being related to α also has great independent usefulness: the electric permeability ε can be defined as the ratio of the capacitance of a capacitor filled with the given insulator to the capacitance of the capacitor in the absence of the insulator. In view of this fact, everywhere below in this chapter we will use the relations for the total (i.e., referring to the whole volume of the insulator) or volume-specific value, employing the electric permeability ε instead of the electric susceptibility α. In this system, using (4.8), Eq. (4.3) takes the form

$$P = \frac{\varepsilon - 1}{4\pi} E. \tag{4.10}$$

It is clear that Eqs. (4.3) and (4.10) can be regarded as equations of state for an insulator in an electric field.

There is no difficulty in transforming from volume-specific to weight-specific quantities.

The electric permeability ε is substantially different for different insulators. It is close to unity for gaseous insulators (1.00007 to 1.015). For nonpolar liquids ε = 2 to 2.5, while for polar liquids (such as water and alcohol) ε has higher values (ε ≈ 80 for water). The values of ε for solid insulators are usually not large (2.5 to 15).

As noted above, for the majority of insulators the electric susceptibility (and consequently also the electric permeability ε) does not depend on the electric field strength E and decreases with increasing temperature.

The nature of the drop in ε with increasing temperature is different for different insulators. Thus, for example, Fig. 4.1 shows the temperature dependence of the electric permeability for water. For some materials (in particular for gases with

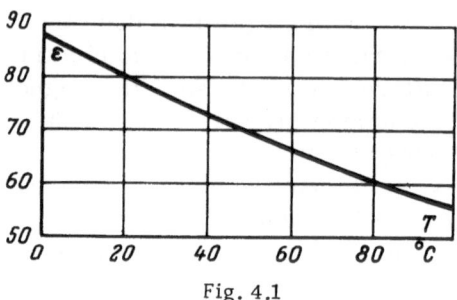

Fig. 4.1

polar molecules) the relation between ε and T can be satisfactorily
described by a relation of the type

$$\varepsilon = 1 + \frac{a}{T},$$ (4.11)

where *a* is a constant. As one can easily see in view of (4.8), this
relation is similar to the Curie law (3.13) for paramagnets. By
analogy with ideal paramagnets, insulators which rigorously obey
relation (4.11) will be called ideal insulators.

The so-called ferroelectrics form a special class of insulators. The electric
permeability of a ferroelectric changes strongly with a change of the external field
strength E. The electric permeability of a ferroelectric reaches huge values in comparison to ordinary insulators.

The special properties of ferroelectric materials are determined by their structure; instead of individual polarized molecules, ferroelectric crystals contain whole
regions of homogeneous polarization (domains).

It is not hard to see than in a certain sense a ferroelectric occupies the same
place among insulators as ferromagnets occupy among magnets.

Ferroelectrics possess residual polarization; after the action of the external
electric field ceases the ferroelectric remains polarized for some time.

Domains exist in ferroelectric materials for temperatures not exceeding a
certain temperature θ_c, the temperature of the so-called ferroelectric Curie point.
At θ_c a ferroelectric crystal undergoes a second-order phase transition. The crystal
goes from the ferroelectric state into the usual insulating state. With T > θ_c the
ferroelectric behaves as an ordinary insulator, and the temperature dependence of its
electric permeability is given by

$$\varepsilon = 1 + \frac{a}{T - \theta_c}$$ (4.12)

in this temperature range.

It is of interest to note that with $T < \theta_C$ the electric susceptibility of a ferro-electric increases with increasing temperature. A graph of the temperature dependence of the electric susceptibility[†] ε for Rochelle salt ($C_4H_4O_6KNa \cdot 4H_2O$) is shown in Fig. 4.2. We note that the Curie temperature for Rochelle salt is 25°C. The maximum in the $\varepsilon(T)$ curve in Fig. 4.2 lies somewhat to the left of the Curie point; evidently the character of the relation $\varepsilon(T)$ for ferroelectrics is analogous to some degree to the relation $\mu(T)$ for a ferromagnet (the Hopkinson effect, see Fig. 3.2).

Figure 4.3 shows the nature of the relations $P(E)$, $D(E)$, and $\varepsilon(E)$ for ordinary insulators, ferroelectrics, and for a vacuum.

4.2. The Basic Thermodynamic Relations for Insulators

It is known that the elementary work performed by an electric field of strength E as the polarization of the insulator increases from P to P + dP is

$$dl^*_V = -EdP, \tag{4.13}$$

or for the insulator as a whole

$$dL^* = -Ed\mathfrak{P}, \tag{4.14}$$

so that the work done on the insulator with increasing polarization has a minus sign in these equations.[‡]

Consequently for an insulator in an electric field the generalized force ξ is the electric field strength E and the generalized coordinate X is the polarization $\mathfrak{P}$ of the insulator.

The combined equation for the first and second laws of thermodynamics (1.28a) for this system as a whole is written in the form

$$TdS = dU + pdV - Ed\mathfrak{P}, \tag{4.15}$$

[†]If the relation $P(E)$ is nonlinear then we introduce the differential quantity $\varepsilon = 1 + 4\pi(P/E)$ instead of the defining equations (4.3) and (4.8) for the electric permeability $\varepsilon = 1 + 4\pi(\partial P/\partial E)_T$.

[‡]By analogy with the scheme used in Chapter 3 for magnets, we do not consider the work done in increasing the strength of the electric field from E to E + dE in the absence of an insulator (in vacuum).

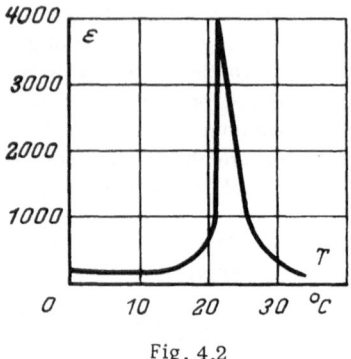

Fig. 4.2

while for volume-specific values it takes the form

$$Tds_v = du_v + pdV - EdP. \tag{4.16}$$

The total enthalpy I^* of an insulator in an electric field, in agreement with (2.66), is given by the relation

$$I^* = U + pV - E\mathfrak{P} \tag{4.17}$$

or, what is the same thing,

$$I^* = I - E\mathfrak{P}. \tag{4.18}$$

Equation (4.15), using (4.17), assumes the form

$$TdS = dI^* - Vdp + \mathfrak{P}dE, \tag{4.19}$$

while for volume-specific quantities

$$Tds_v = di^*_v - dp + PdE. \tag{4.20}$$

Furthermore, using (4.10) and (4.7), Eq. (4.14) for the elementary work dL^* can be written as follows:

$$dL^* = -\frac{\varepsilon - 1}{4\pi} EdE = -\frac{\varepsilon - 1}{4\pi\varepsilon} EdD. \tag{4.21}$$

The Maxwell equations (2.144a), (2.147a), (2.150a), and (2.153a) for the insulator as a whole and for the electric field assume the following forms:

$$\left(\frac{\partial \mathfrak{B}}{\partial T}\right)_{S,V} = \left(\frac{\partial S}{\partial E}\right)_{\mathfrak{B},V}, \qquad (4.22)$$

$$\left(\frac{\partial \mathfrak{B}}{\partial S}\right)_{E,p} = -\left(\frac{\partial T}{\partial E}\right)_{S,p}, \qquad (4.23)$$

$$\left(\frac{\partial \mathfrak{B}}{\partial S}\right)_{T,V} = -\left(\frac{\partial T}{\partial E}\right)_{\mathfrak{B},V}, \qquad (4.24)$$

$$\left(\frac{\partial \mathfrak{B}}{\partial T}\right)_{E,p} = \left(\frac{\partial S}{\partial E}\right)_{T,p}. \qquad (4.25)$$

The Maxwell equations for the volume-specific quantities have the same form with the only difference being that the values of $\mathfrak{B}$ and S are replaced by P and s_V, respectively.

The relations for the dependence of caloric properties (u_V and i_V^*) of an insulator on the electric field strength E and the polarization P can be found from (4.16) and (4.20) using the Maxwell equations.

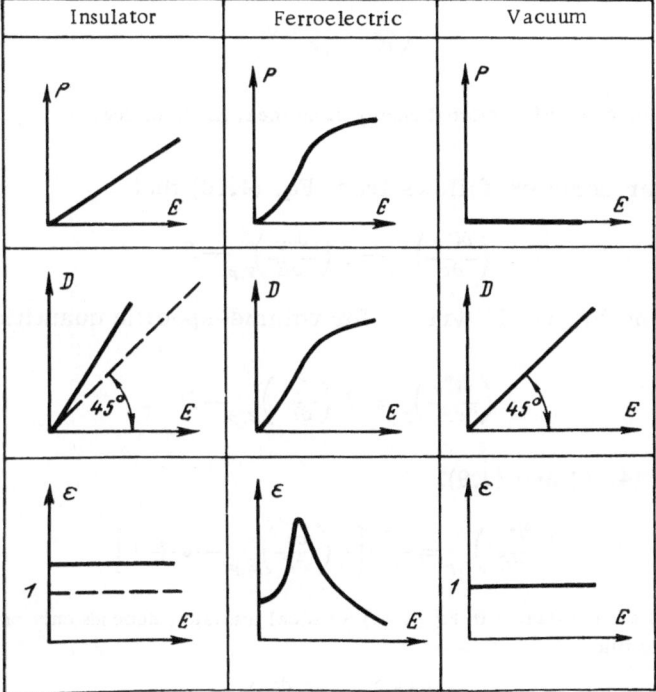

Fig. 4.3

It follows from Eq. (4.16) that

$$\left(\frac{\partial u_V}{\partial P}\right)_{T,V} = T\left(\frac{\partial s_V}{\partial P}\right)_{T,V} + E. \tag{4.26}$$

Using Eq. (4.24) written for volume-specific quantities, we find

$$\left(\frac{\partial u_V}{\partial P}\right)_{T,V} = E - T\left(\frac{\partial E}{\partial T}\right)_{P,V} \tag{4.27}$$

or, using (4.10),

$$\left(\frac{\partial u_V}{\partial P}\right)_{T,V} = E\left[1 + \frac{T}{\varepsilon - 1}\left(\frac{\partial \varepsilon}{\partial T}\right)_{P,V}\right]. \tag{4.28}$$

For an ideal insulator, in agreement with (4.11)

$$\left(\frac{\partial \varepsilon}{\partial T}\right)_{P,V} = -\frac{a}{T^2} = -\frac{\varepsilon - 1}{T} \tag{4.29}$$

and consequently

$$\left(\frac{\partial u_V}{\partial P}\right)_{T,V} = 0, \tag{4.30}$$

i.e., the volume-specific internal energy of an ideal insulator does not depend on the polarization.

It furthermore follows from Eq. (4.20) that

$$\left(\frac{\partial i^*_V}{\partial E}\right)_{T,p} = T\left(\frac{\partial s_V}{\partial E}\right)_{T,p} - P. \tag{4.31}$$

Using Eq. (4.25) written for volume-specific quantities, we find

$$\left(\frac{\partial i^*_V}{\partial E}\right)_{T,p} = T\left(\frac{\partial P}{\partial T}\right)_{E,p} - P, \tag{4.32}$$

or, using (4.10) and (4.9),

$$\left(\frac{\partial i^*_V}{\partial E}\right)_{T,p} = \frac{E}{4\pi}\left[T\left(\frac{\partial \varepsilon}{\partial T}\right)_{E,p} - \varepsilon + 1\right]. \tag{4.33}$$

Since the constant a in Eq. (4.11) for ideal insulators depends only on the temperature, we find

$$\left(\frac{\partial \varepsilon}{\partial T}\right)_{E,p} = \left(\frac{\partial \varepsilon}{\partial T}\right)_{P,V}, \tag{4.34}$$

and using (4.29) and (4.33) we find the obvious result

$$\left(\frac{\partial i^*_V}{\partial E}\right)_{T,\,p} = -\frac{\varepsilon-1}{2\pi}E\,, \tag{4.35}$$

i.e., with increasing electric field strength the enthalpy of an ideal insulator decreases in proportion to E.

4.3. The Heat Capacities of an Insulator

By analogy with the heat capacities c_p and c_v for "ordinary" systems and the heat capacities $c_{H,p}$ and $c_{j,p}$ for magnets, for an insulator we introduce the idea of the heat capacities at constant external field strength, c_E, and at constant polarization c_p. If these heat capacities are treated under conditions of constant pressure in the medium in which the insulator is placed, they are denoted, respectively, by $c_{E,p}$ and $c_{P,p}$.

The volume-specific values of these heat capacities will be denoted by $C_{E,p}$ and $C_{P,p}$, respectively. Evidently

$$c_{E,p} = C_{E,p}v \quad \text{and} \quad c_{P,p} = C_{P,p}v, \tag{4.36}$$

where v is the specific volume of the insulator.

It is further evident that

$$C_{E,p} = T\left(\frac{\partial s_v}{\partial T}\right)_{E,p} \text{ and } C_{P,p} = T\left(\frac{\partial s_v}{\partial T}\right)_{P,p}. \tag{4.37}$$

In agreement with Eq. (1.49), we can write

$$\left(\frac{\partial s_v}{\partial T}\right)_{E,p} = \left(\frac{\partial s_v}{\partial T}\right)_{P,p} + \left(\frac{\partial s_v}{\partial P}\right)_{T,p}\left(\frac{\partial P}{\partial T}\right)_{E,p}. \tag{4.38}$$

It is further evident that

$$\left(\frac{\partial s_v}{\partial P}\right)_{T,p} = \left(\frac{\partial s_v}{\partial E}\right)_{T,p}\left(\frac{\partial E}{\partial P}\right)_{T,p}. \tag{4.39}$$

Using this relation and Eq. (4.25) written for volume-specific quantities, we find from (4.38) that

$$C_{E,p} - C_{P,p} = T\left(\frac{\partial P}{\partial T}\right)_{E,p}^2\left(\frac{\partial E}{\partial P}\right)_{T,p}. \tag{4.40}$$

This relation connects the values of $C_{E,p}$ and $C_{P,p}$ for an insulator. It is not hard to see that this relation is analogous to Eq. (3.59) for the heat capacities c_p and c_v of "ordinary" systems and corresponds to Eq. (3.60) for the heat capacities $c_{H,p}$ and $c_{j,p}$ of a magnet.

Using the obvious relation

$$\left(\frac{\partial P}{\partial T}\right)_E \left(\frac{\partial T}{\partial E}\right)_P \left(\frac{\partial E}{\partial P}\right)_T = -1 \tag{4.41}$$

Eq. (4.40) can also be put in the form

$$C_{E,p} - C_{P,p} = -T\left(\frac{\partial P}{\partial T}\right)_{E,p} \left(\frac{\partial E}{\partial T}\right)_{P,p} \tag{4.42}$$

or in the form

$$C_{E,p} - C_{P,p} = T\left(\frac{\partial E}{\partial T}\right)_{P,p}^2 \left(\frac{\partial P}{\partial E}\right)_{T,p}. \tag{4.43}$$

The equation for the difference $C_{E,p} - C_{P,p}$ can also be obtained in a different way using Eq. (2.156).

The partial derivatives which enter Eqs. (4.40), (4.42), and (4.43), can be expressed as follows using Eq. (4.10):

$$\left(\frac{\partial P}{\partial T}\right)_{E,p} = \frac{E}{4\pi}\left(\frac{\partial \epsilon}{\partial T}\right)_{E,p}, \tag{4.44}$$

$$\left(\frac{\partial P}{\partial E}\right)_{T,p} = \frac{1}{4\pi}\left[E\left(\frac{\partial \epsilon}{\partial E}\right)_{T,p} + \epsilon - 1\right], \tag{4.45}$$

and

$$\left(\frac{\partial E}{\partial T}\right)_{P,p} = -\frac{E}{\epsilon - 1}\left(\frac{\partial \epsilon}{\partial T}\right)_{P,p}. \tag{4.46}$$

If we have data on the dependence of the permeability ϵ of the insulator on the temperature and the electric field strength E we can use them to calculate the difference between the heat capacities $C_{E,p}$ and $C_{P,p}$. For most insulators with values of E that are not too high we have

$$\left(\frac{\partial \epsilon}{\partial E}\right)_{T,\nu} = 0 \tag{4.47}$$

and consequently

$$\left(\frac{\partial P}{\partial E}\right)_{T,p} = \frac{\varepsilon - 1}{4\pi} \cdot \tag{4.48}$$

Ferroelectric materials are an exception to this rule.

For an ideal insulator, using (4.11), we find from Eq. (4.40) that

$$C_{E,p} - C_{P,p} = \frac{\varepsilon - 1}{4\pi} \frac{E^2}{T} \tag{4.49}$$

or

$$C_{E,p} - C_{P,p} = \frac{EP}{T}. \tag{4.50}$$

Like Eq. (3.81) for magnets, Eq. (4.50) is to some degree analogous to formula (3.83) for an ideal gas.

It is further obvious from Eq. (4.37) that

$$\left(\frac{\partial C_{E,p}}{\partial E}\right)_{T,p} = T\left[\frac{\partial}{\partial E}\left(\frac{\partial s_v}{\partial T}\right)_{E.}\right]_{T,p} \tag{4.51}$$

or, what is the same,

$$\left(\frac{\partial C_{E,p}}{\partial E}\right)_{T,p} = T\left[\frac{\partial}{\partial T}\left(\frac{\partial s_v}{\partial E}\right)_{T,p}\right]_{E,p}. \tag{4.52}$$

Using (4.25) we find

$$\left(\frac{\partial C_{E,p}}{\partial E}\right)_{T,p} = T\left(\frac{\partial^2 P}{\partial T^2}\right)_{E,p}, \tag{4.53}$$

an equation relating the dependence of the isothermal heat capacity $C_{E,p}$ (with p = const) on the electric field strength E to the derivative $(\partial^2 P/\partial T^2)_{E,p}$. Integrating, we find from (4.53) that

$$C_{E,p}(E,T,p) - C_{E,p}(E=0,T,p) = T\int_0^E \left(\frac{\partial^2 P}{\partial T^2}\right)_{E,p} dE, \tag{4.54}$$

where the integral on the right-hand side is calculated along an isotherm T = const (with p = const).

Further,

$$C_{E,p}(E=0,T,p) = C_p(T,p). \tag{4.55}$$

Consequently Eq. (4.54) can be written in the form

$$C_{E,p}(E,T,p) - C_p(T,p) = T \int_0^E \left(\frac{\partial^2 P}{\partial T^2} \right)_{E,p} dE. \qquad (4.56)$$

Equation (4.56) shows how the isobaric heat capacity of an insulator changes as it is subjected to an electric field of strength E.

Using (4.10) we can represent this equation as follows:

$$C_{E,p}(E,T,p) - C_p(T,p) = \frac{T}{4\pi} \int_0^E \left(\frac{\partial^2 \varepsilon}{\partial T^2} \right)_{E,p} E\,dE. \qquad (4.57)$$

Using (4.11) we find the following equation for ideal insulators:

$$C_{E,p} - C_p = \frac{EP}{T}. \qquad (4.58)$$

From a comparison of this equation with Eq. (4.50) it is evident that for an ideal insulator

$$C_{P,p} = C_p.$$

In ending the discussion of the heat capacities of an insulator we mention that the heat capacity of a ferroelectric passes through a maximum at the ferroelectric Curie point, which is characteristic of a second-order phase transformation.

4.4. Thermodynamic Processes in Insulators

The characteristics of the basic reversible thermodynamic processes in insulators can be obtained by methods similar to those used in Section 3.4 in describing thermodynamic processes in magnets. For the sake of brevity we will omit certain intermediate calculations in the ensuing paragraphs.

Isothermal Processes. The change in the volume-specific entropy of an insulator in an isothermal process (with p = const also) is given by the obvious relation

$$s_{v_2}(E_2,T) - s_{v_1}(E_1,T) = \int_{E_1}^{E_2} \left(\frac{\partial s_v}{\partial E} \right)_{T,p} dE, \qquad (4.59)$$

which can be represented in the form

$$s_{v_2}(E_2, T) - s_{v_1}(E_1, T) = \int_{E_1}^{E_2} \left(\frac{\partial P}{\partial T} \right)_{E,P} dE \qquad (4.60)$$

using Eq. (4.25), or, what is the same,

$$s_{v_2}(E_2, T) - s_{v_1}(E_1, T) = \frac{1}{4\pi} \int_{E_1}^{E_2} \left(\frac{\partial \varepsilon}{\partial T} \right)_{E,P} dE. \qquad (4.61)$$

Since, as noted previously, the polarization in an insulator decreases with increasing temperature, it is obvious from these relations that the volume-specific entropy of an insulator decreases along an isotherm with increasing electric field strength. An exception to this rule is a ferroelectric for temperatures $T < \Theta_c$ i.e., in the temperature range where $(\partial \varepsilon / \partial T)_{E,p} > 0$ (see Fig. 4.2).

For the particular case of an ideal insulator we find from (4.61), using Eq. (4.11), that

$$s_{v_2}(E_2, T) - s_{v_1}(E_1, T) = - \frac{a(E_2^2 - E_1^2)}{8\pi T^2}. \qquad (4.62)$$

The amount of heat supplied to a unit volume of the insulator (or flowing out of it) in an isothermal process is determined from Eq. (3.101):

$$q_{v_{1-2}} = T(s_{v_2} - s_{v_1}),$$

where $(s_{v_2} - s_{v_1})$ is the change in the volume-specific entropy in an isothermal process described by Eqs. (4.60) and (4.61). Since we have $s_{v_2} < s_{v_1}$, in an isothermal process in an insulator, as shown above, it follows that when an insulator is isothermally polarized (i.e., with increasing electric field strength E) it liberates heat $(q_{v_{1-2}} < 0)$ while with decreasing electric field strength it absorbs heat $(q_{v_{1-2}} > 0)$.

The work done by the electric field in polarizing the insulator in an isothermal process is determined using Eq. (4.13), which shows that for a unit volume of the insulator we have

$$l^*_{v_{1-2}} = - \int_{P_1}^{P_2} E \, dP \qquad (4.63)$$

or, if we take (4.10) into account,

$$l^*_{v_{1-2}} = -4\pi \int_{P_1}^{P_2} \frac{P}{\varepsilon - 1} dP, \qquad (4.64)$$

where the integration runs along an isotherm.

For a normal insulator in which the electric susceptibility does not change with E or P, it follows that

$$l^*_{v_{1-2}} = -2\pi \frac{P_2^2 - P_1^2}{\varepsilon - 1} \qquad (4.65)$$

or

$$l^*_{v_{1-2}} = -\frac{E_2 P_2 - E_1 P_1}{2}. \qquad (4.66)$$

For ferroelectric materials, $l^*_{v_{1-2}}$ is calculated directly from Eqs. (4.63) or (4.64) using data on the relation between the electric permeability and the electric field strength E at a given temperature.

Adiabatic Processes. As we know, in a reversible adiabatic process

$$s = \text{const}$$

and

$$q_{1-2} = 0.$$

The work done by the electric field in polarizing an insulator adiabatically is given by Eqs. (4.63)-(4.66); however, in this case the integrals in Eqs. (4.63)-(4.64) are calculated along an isentrope rather than an isotherm.

Processes with E = const (with p = const). The amount of heat supplied to the insulator (or liberated by it) in this process is given by

$$q_{v_{1-2}} = i^*_{v_2} - i^*_{v_1}. \qquad (4.67)$$

The change in the volume-specific entropy of the insulator in this process is given by

$$s_{v_2}(E,T_2) - s_{v_1}(E,T_1) = \int_{T_1}^{T_2} \left(\frac{\partial s_v}{\partial T}\right)_{E,p} dT, \tag{4.68}$$

whence, using Eq. (4.37), we find

$$s_{v_2}(E,T_2) - s_{v_1}(E,T_1) = \int_{T_1}^{T_2} \frac{C_{E,p}}{T} dT. \tag{4.69}$$

The work in this process is determined from Eq. (4.63) with the integration done along a line E = const,

$$l^*_{v_{1-2}} = -E(P_2 - P_1) \tag{4.70}$$

or, using Eq. (4.10),

$$l^*_{v_{1-2}} = -\frac{E^2}{4\pi}[\varepsilon(T_2) - \varepsilon(T_1)]. \tag{4.71}$$

Since the electric permeability decreases with increasing temperature for a normal insulator, when $T_2 > T_1$ we find from (4.71) that $l^*_{v_{1-2}} > 0$, i.e., work is done by the insulator.

For ferroelectrics with $T < \theta_C$ we find $l^*_{v_{1-2}} < 0$ near θ_C.

For an ideal insulator we find from (4.71), using (4.11), that

$$l^*_{v_{1-2}} = \frac{aE^2}{4\pi}\left(\frac{1}{T_1} - \frac{1}{T_2}\right). \tag{4.72}$$

Processes with P = const (and p = const). The amount of heat supplied to the insulator (or liberated by it) in a process with P = const (and p = const) is given by

$$q_{v_{1-2}} = i_{v_2} - i_{v_1}, \tag{4.73}$$

where i_v is the "ordinary" volume-specific enthalpy $i_v = u_v + p$.

The change in the volume-specific entropy of an insulator in this process is given by

$$s_{v_2}(P,T_2) - s_{v_1}(P,T_1) = \int_{T_1}^{T_2} \left(\frac{\partial s_v}{\partial T}\right)_{P,p} dT \tag{4.74}$$

or, using (4.37),

$$s_{v_2}(P,T_2) - s_{v_1}(P,T_1) = \int_{T_1}^{T_2} \frac{C_{P,p}}{T}\, dT.\tag{4.75}$$

Since the polarization of an insulator remains constant in this process, we see from (4.63) that for this process

$$l^*_{v_{1-2}} = 0.\tag{4.76}$$

4.5. The Piezoelectric, Electrostriction, Electrocaloric, and Pyroelectric Effects

The change in the polarization of an insulator with a change in the external pressure is called the piezoelectric effect.

The piezoelectric effect was discovered in 1880 by the Curie brothers, who observed the appearance of electric charge on the surface of a quartz plate when it was deformed; a compressive force of 1 kg/cm^2 caused the appearance of a potential difference of hundredths of a volt on opposite faces of the quartz crystal. This effect appears in many crystals and is particularly large in ferroelectric crystals. Of the piezoelectric materials quartz is the most extensively used in practical applications; it has a smaller piezoelectric effect than Rochelle salt but has greater mechanical and electrical strength.

The piezoelectric effect is widely used in modern technology. Piezoelectric transducers are used in various equipment for transforming mechanical action in an insulator into an electrical signal. This kind of transducer is used in piezoelectric manometers, in tension-measuring devices, in accelerometers, in piezoelectric transducers (for transforming sound waves into an electric current, for example, in microphones), and in many other kinds of piezoelectric transducers.

For insulators in general the piezoelectric effect can be characterized by the magnitude of the derivative $(\partial\mathfrak{P}/\partial p)$.

It should be kept in mind that the amount of polarization will, strictly speaking, differ depending on whether the pressure is changed isothermally or adiabatically.

The relation between $(\partial\mathfrak{P}/\partial p)_{S,E}$ and $(\partial\mathfrak{P}/\partial p)_{T,E}$ is given by an equation which follows from Eq. (1.49):

$$\left(\frac{\partial\mathfrak{P}}{\partial p}\right)_{S,E} = \left(\frac{\partial\mathfrak{P}}{\partial p}\right)_{T,E} + \left(\frac{\partial\mathfrak{P}}{\partial T}\right)_{p,E}\left(\frac{\partial T}{\partial p}\right)_{S,E}.\tag{4.77}$$

Here $(\partial T/\partial p)_{S,E}$ is a quantity which characterizes the change in the temperature of the insulator during adiabatic compression (under conditions where E = const). It is clear that

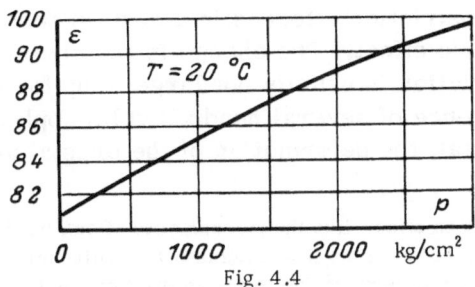

Fig. 4.4

$$\left(\frac{\partial T}{\partial p}\right)_{S,E} = -\left(\frac{\partial T}{\partial S}\right)_{p,E}\left(\frac{\partial S}{\partial p}\right)_{T,E}, \tag{4.78}$$

whence, using Eqs. (4.37) and (2.154a), we find that

$$\left(\frac{\partial T}{\partial p}\right)_{S,E} = \frac{T}{C_{E,p}V}\left(\frac{\partial V}{\partial T}\right)_{p,E}; \tag{4.79}$$

where $C_{E,p}$ is the volume-specific heat capacity and $C_{E,p}V$ is the total heat capacity of the insulator.

In agreement with Eq. (4.9) the derivative $(\partial \mathfrak{P}/\partial p)_E$ can be represented as

$$\left(\frac{\partial \mathfrak{P}}{\partial p}\right)_E = V\left(\frac{\partial P}{\partial p}\right)_E + P\left(\frac{\partial V}{\partial p}\right)_E, \tag{4.80}$$

and using (4.10) we find

$$\left(\frac{\partial \mathfrak{P}}{\partial p}\right)_E = \frac{VE}{4\pi}\left[\left(\frac{\partial \varepsilon}{\partial p}\right)_E + (\varepsilon - 1)\frac{1}{V}\left(\frac{\partial V}{\partial p}\right)_E\right]. \tag{4.81}$$

The quantity $\frac{1}{V}(\partial V/\partial p)_E$ which enters this equation is the compressibility of the insulator; if the piezoelectric effect is treated under adiabatic conditions then this is the adiabatic compressibility, while under isothermal conditions it is the isothermal compressibility.

The quantity $(\partial \varepsilon/\partial p)_E$ characterizes the pressure dependence of the electric permeability. This relation is different for different insulators. As an example, Fig. 4.4 shows the relation between ε and p for water.

By the e l e c t r o s t r i c t i o n e f f e c t we mean a change in the dimensions of an insulator with a change in the electric field strength E.

The electrostriction effect is in a certain sense the opposite of the piezoelectric effect. The change in the size of an insulator due to electrostriction is usually not large; thus for quartz, with a potential difference of several hundred volts applied to opposite faces of the crystal, the deformation of the crystal amounts to about 10^{-6} mm.

The electrostriction effect, like the piezoelectric effect, has found various practical applications. For example if an alternating electric field acts on a piezo-electric the change in its dimensions due to electrostriction can be used to generate mechanical vibrations of the desired frequency. There is particular interest in the case where the vibration frequency of the applied voltage coincides with an intrinsic vibration frequency of the piezoelectric material, so that the amplitude of the vibrations is particularly large; such a structure may be used to generate ultrasonic vibrations. Various piezoelectric generators using the electrostriction effect are widely used in radio technology and other areas of technology. In particular we mention piezoelectric quartz frequency stabilizing elements for electromagnetic oscillation generators and the use of this effect in piezoelectric quartz as a frequency standard (quartz clocks).

The electrostriction effect in an isotropic insulator with a constant external pressure p is characterized by the derivative $(\partial V/\partial E)$. Depending on the conditions under which E is changed — isothermal or adiabatic — $(\partial V/\partial E)$ will be different, as follows from (1.49):

$$\left(\frac{\partial V}{\partial E}\right)_{S,p} = \left(\frac{\partial V}{\partial E}\right)_{T,p} + \left(\frac{\partial V}{\partial T}\right)_{p,E}\left(\frac{\partial T}{\partial E}\right)_{S,p}. \qquad (4.82)$$

Here $(\partial V/\partial T)_{p,E}$ is the ordinary thermal expansion of the insulator; the meaning of $(\partial T/\partial E)_{S,p}$ will be explained below.

It is obvious from general thermodynamic considerations that there should be a unique relation between the piezoelectric and electrostriction effects.

The relations between $(\partial \mathfrak{P}/\partial p)_{T,E}$ and $(\partial V/\partial E)_{T,p}$ and $(\partial \mathfrak{P}/\partial p)_{S,E}$ and $(\partial V/\partial E)_{T,p}$ can be obtained as follows.

In agreement with Eqs. (2.47) and (2.50) the expression for the isobaric—isothermal potential of the systems treated in this chapter is written in the form

$$\Phi^* = U - TS + pV - E\mathfrak{P}. \qquad (4.83)$$

Using (4.15) it follows that

$$d\Phi^* = -SdT + Vdp - \mathfrak{P}dE. \tag{4.84}$$

The quantity Φ^* is a state function and consequently its differential is a total differential. It is obvious from this that

$$V = \left(\frac{\partial \Phi^*}{\partial p}\right)_{T,E}, \quad \mathfrak{P} = -\left(\frac{\partial \Phi^*}{\partial E}\right)_{T,p} \tag{4.85}$$

In agreement with (1.43) we find

$$\left(\frac{\partial V}{\partial E}\right)_{T,p} = -\left(\frac{\partial \mathfrak{P}}{\partial p}\right)_{T,E}. \tag{4.86}$$

Using Eq. (4.81) we find

$$\left(\frac{\partial V}{\partial E}\right)_{T,p} = -\frac{VE}{4\pi}\left[\left(\frac{\partial \varepsilon}{\partial p}\right)_{T,E} + (\varepsilon - 1)\frac{1}{V}\left(\frac{\partial V}{\partial p}\right)_{T,E}\right]. \tag{4.87}$$

In an analogous way we find from Eqs. (4.17) and (4.19) that

$$dI^* = TdS + Vdp - \mathfrak{P}dE, \tag{4.88}$$

whence it is obvious that

$$V = \left(\frac{\partial I^*}{\partial p}\right)_{S,E}, \quad \mathfrak{P} = -\left(\frac{\partial I^*}{\partial E}\right)_{S,p} \tag{4.89}$$

and consequently

$$\left(\frac{\partial V}{\partial E}\right)_{S,p} = -\left(\frac{\partial \mathfrak{P}}{\partial p}\right)_{S,E} \tag{4.90}$$

or, using (4.81),

$$\left(\frac{\partial V}{\partial E}\right)_{S,p} = -\frac{VE}{4\pi}\left[\left(\frac{\partial \varepsilon}{\partial p}\right)_{S,E} + (\varepsilon - 1)\frac{1}{V}\left(\frac{\partial V}{\partial p}\right)_{S,E}\right]. \tag{4.91}$$

The change in the polarization of an insulator with a change in its temperature is called the pyroelectric effect. The

effect was first observed in 1703 when it was found that heating
a prismatic tourmaline crystal causes electric charge to appear
on certain faces of the crystal. An effect opposite to the pyroelectric effect, the so-called e l e c t r o c a l o r i c   e f f e c t, was discovered experimentally (the change in the temperature of an insulator with a change in the electric field strength).

Evidently the pyroelectric effect can be characterized by
the derivative $(\partial P/\partial T)$ and correspondingly the electrocaloric effect
is characterized by the derivative $(\partial T/\partial P)$.

In principle the electrocaloric effect will occur in any thermodynamic process which occurs in an insulator (other than isothermal processes). The greatest practical interest lies in adiabatic processes, which occur with a rapid change in the electric
field strength.

The change in the temperature of an insulator with a change
of E in an adiabatic process between states 1 and 2 is given by
the obvious relation

$$T_2 - T_1 = \int\limits_{E_1}^{E_2} \left(\frac{\partial T}{\partial E}\right)_{s_V, p} dE. \tag{4.92}$$

Since

$$\left(\frac{\partial T}{\partial E}\right)_{s_V, p} = - \left(\frac{\partial s_V}{\partial E}\right)_{T, p} \left(\frac{\partial T}{\partial s_V}\right)_{E, p}, \tag{4.93}$$

using Eqs. (5.25) and (5.37), we have

$$\left(\frac{\partial T}{\partial E}\right)_{s_V, p} = - \frac{T}{C_{E, p}} \left(\frac{\partial P}{\partial T}\right)_{E, p} \tag{4.94}$$

(here s_V and $C_{E,p}$ are the volume-specific entropy and heat capacity).

Since the heat capacity $C_{E,p}$ of an insulator is positive, while
$(\partial P/\partial T)_{E,p}$ is negative (the polarization decreases with increasing
temperature), its temperature increases with increasing electric
field strength (in an adiabatic process with p = const).

Using Eq. (4.94) we find from (4.92) that

$$T_2 - T_1 = - \int\limits_{E_1}^{E_2} \frac{T}{C_{E, p}} \left(\frac{\partial P}{\partial T}\right)_{E, p} dE \tag{4.95}$$

or, with (4.10) in mind,

$$T_2 - T_1 = -\frac{1}{4\pi} \int_{E_1}^{E_2} \frac{T}{C_{E,p}} \left(\frac{\partial \varepsilon}{\partial T}\right)_{E,p} E\, dE. \tag{4.96}$$

The electrocaloric effect is usually quite small. With this in mind Eq. (4.96) can be written as

$$T_2 - T_1 = -\frac{T_{av}\,(E_2^2 - E_1^2)}{8\pi C_{E,p}} \left(\frac{\partial \varepsilon}{\partial T}\right)_{E,p}, \tag{4.97}$$

where $T_{av} = (T_1 + T_2)/2$ while $C_{E,p}$ and $(\partial\varepsilon/\partial T)_{E,p}$ are values averaged over the given temperature and electric field ranges.

CHAPTER 5

Superconductivity

5.1. Introduction

By superconductivity we mean a special physical state of
certain pure metals and alloys in which the electrical resistance
of the metal (or alloy) is zero. Materials in which the superconducting state is observed are called superconductors. The phenomenon of superconductivity was first found in 1911 by H. Kammer-
lingh-Onnes, who was studying the temperature dependence of the
electrical resistance of mercury at temperatures close to absolute
zero.

The transformation of a superconductor from the normal
state (which has a definite value of the electrical resistance) into
the superconducting state occurs when it is cooled below a certain

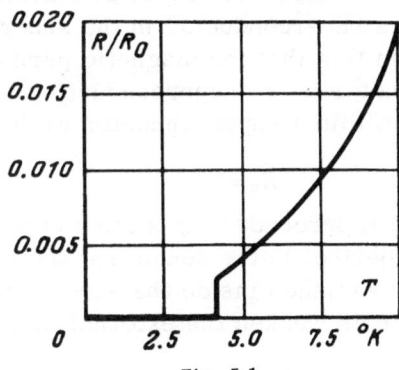

Fig. 5.1

119

temperature, called the critical temperature of the particular superconductor (we denote it by T_c). The critical temperature has different values for different superconductors. The critical temperatures of the known superconductors lie between 0.012°K (for tungsten) and 20.05°K (for a solid solution of Nb_3Al and Nb_3Ge). At the temperature T_c the electrical resistance of the superconductor decreases discontinuously from a finite value to zero (see Fig. 5.1, which shows the temperature dependence of R/R_0 for mercury; here R and R_0 are the resistance of mercury at the given temperature and at 0°C).

It should be noted that the resistance of a superconductor is not simply very small but is rigorously zero in the superconducting state:

$$R = 0. \tag{5.1}$$

In a closed circuit made of a superconducting material and cooled below T_c, current can circulate an arbitrarily long time without decaying.

Since the resistance of a superconductor is zero with $T < T_c$, it is evident that there is no Joule energy loss (I^2R) when a current I passes through a superconductor. It is clear from this that in principle very large currents can pass through a superconducting conductor. This remarkable property of superconductors is of great promise for their widespread use in technology.

Zero electrical resistance is the most important property of the superconducting state. A second fundamental feature of superconductivity was found by W. Meissner and R. Ochsenfeld in 1933, when they experimentally discovered that an external magnetic field cannot penetrate into a superconductor in the superconducting state. We may conclude from this that the magnetic permeability μ is zero for a superconductor in the superconducting state so that in agreement with (3.7) within a superconductor we have†

$$B_c = 0; \tag{5.2}$$

in other words, in the superconducting state a superconductor is a perfect diamagnet (perfect in the sense that the external magnetic field completely vanishes inside the superconductor; we recall that all diamagnets weaken the external magnetic field).

---

†Everywhere below we will consider only the so-called classical superconductors (type I superconductors). The so-called type II superconductors have behavioral rules which are quite complicated and at the present time are still not sufficiently studied, and so are not considered here.

In agreement with (3.5) for all magnetic materials we have

$$B = H + 4\pi J,$$

so that in view of (5.2) we find for a superconductor in the super-conducting state

$$J_s = -\frac{1}{4\pi} H \qquad (5.3)$$

(note that in what follows quantities referring to the superconductor in the normal and superconducting states will be denoted by the subscripts n and s, respectively).

We recall that the magnetization J which enters Eq. (3.5) is the volume-specific magnetization. Consequently the quantity J_s defined by Eq. (5.3) is also the magnetization of a unit volume of the superconductor. Obviously in agreement with Eq. (3.9)

$$j = Jv$$

the weight-specific magnetization of a superconductor is written as

$$j_s = -\frac{v_s}{4\pi} H, \qquad (5.4)$$

where v_c is the specific volume of the superconductor in the super-conducting state.

This relation defines the magnetization of a superconductor in the superconducting state.[†]

It follows from this relation that the magnetic susceptibility of a superconductor in the superconducting state is

$$\chi_s = -\frac{v_s}{4\pi}. \qquad (5.5)$$

When the superconductor is in the normal state ($T > T_c$) the magnetic field penetrates into the superconductor; in the normal state a superconductor is an ordinary diamagnet or para-magnet. Here the magnetic susceptibility of the superconductor is quite small since in agreement with (3.10) we can assume with

---

[†]The magnetization of a superconductor is explained as follows. The external magnetic field does not penetrate into the superconductor and drops to zero in a thin (of the order of 10^{-5} to 10^{-6} cm) surface layer of the superconductor in which an undamped supercurrent flows. This current compensates the external magnetic field and does not allow it to penetrate into the sample.

a high degree of precision that the superconductor has no magnetization in the normal state, i.e.,

$$\chi_n = 0 \qquad\qquad (5.6)$$

and

$$J_n = 0. \qquad\qquad (5.7)$$

Consequently, in agreement with (3.5),

$$B_n = H. \qquad\qquad (5.8)$$

If a sufficiently strong external magnetic field is imposed
on a superconductor in the superconducting state (i.e., with $T < T_c$)
superconductivity is destroyed, i.e., the magnetic field penetrates
into the superconductor and it goes into the normal state (despite
the fact that the temperature of the superconductor is less than
T_c). It turns out that the lower the temperature to which the
superconductor is cooled, the greater is the external magnetic field
required to destroy superconductivity (called the critical magnetic
field and denoted by H_c).

The results given below in this chapter refer to an infinitely long superconducting cylinder in a uniform longitudinal magnetic field. For a superconductor of
arbitrary shape, attention must be paid to the fact that the presence of the superconductor distorts the external uniform magnetic field. As a result, as the external
magnetic field strength increases the field strength at the surface of the superconductor will in general be different at different points on the surface (this fact can be
accounted for with the aid of the demagnetizing factor which was introduced previously — see p. 55). Consequently at different points on the sample surface the field
strength reaches H_c for different values of H; this effect causes the so-called intermediate state in the superconductor.

The relation between H_c and the temperature for mercury
is shown in Fig. 5.2. The relation $H_c(T)$ is a property of each particular superconductor. Here H_c can vary over broad limits
which differ by factors of hundreds and thousands for different superconductors. In many cases the relation between H_c and the temperature can be approximated by the following empirical formula
with good precision:

$$H_c = H_0 \left[1 - \left(\frac{T}{T_c} \right)^2 \right], \qquad\qquad (5.9)$$

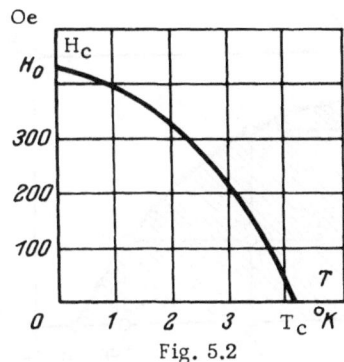

Fig. 5.2

where H_c is the strength of the critical magnetic field at a temperature T, and H_0 and T_c are constants. The meaning of the constants H_0 and T_c which enter this parabolic formula is not hard to establish. Indeed it is evident from (5.9) that $H_c = 0$ only with $T = T_c$. Consequently T_c is the critical temperature of the superconductor in the absence of an external magnetic field. Furthermore if $T = 0$ then $H_c = H_0$ as we see from (5.9). This means that H_0 is the critical magnetic field of the superconductor at absolute zero (it should be noted that since, in agreement with the Nernst law, one cannot reach absolute zero, H_0 is to be found by extrapolating the curve for $H_c(T)$ to the intersection with the ordinate). The values of T_c and H_0 for some superconductors are listed in Table 5.1.

We consider one of the basic state diagrams for a superconductor, the H–T diagram shown in Fig. 5.3. In the H–T diagram the shaded region under the curve $H_c(T)$ corresponds to the superconducting state and the unshaded region over the curve corresponds to the normal state of the superconductor. Thus a superconductor for which this diagram is constructed is in the superconducting

TABLE 5.1

Super-conductor	T_c, °K	H_0, Oe	Super-conductor	T_c, °K	H_0, Oe
Niobium	8	2600	Thallium	2.38	171
Lead	7.22	805	Aluminum	1.20	106
Vanadium	5.1	~1200	Gallium	1.10	50.3
Tantalum	4.4	975	Zinc	0.91	52.5
Mercury	4.152	413	Osmium	0.71	65
Tin	3.730	304.5	Cadmium	0.56	28.8
Indium	3.37	269	Ruthenium	0.47	46

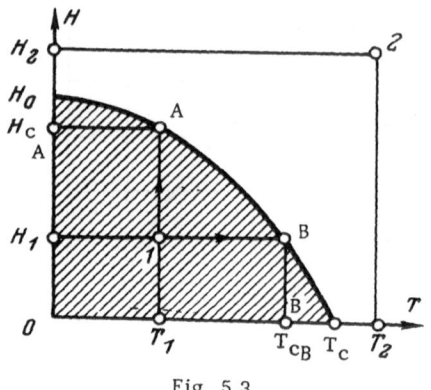

Fig. 5.3

state (point 1) at a field strength H_1 and a temperature T_1, while
for H_2 and T_2 it is in the normal state (point 2). It is further
evident that a superconductor in the superconducting state corresponding to point 1 can be put into the normal state in different ways. For
example, we can go along a path of increasing temperature, holding
the external magnetic field constant (H_1 = const).

When the temperature reaches T_{c_B} [point B is the intersection of the line H_1 = const and the curve $H_c(T)$] the superconductor
goes into the normal state. The transformation of the superconductor from the superconducting to the normal state can also take
place in a "combined" manner — with simultaneous increases in
the temperature and the magnetic field strength.

It should be noted, as shown by experimental data, that the
curve $H_c(T)$ has a horizontal tangent at 0°K, i.e., at this temperature $dH_c/dT = 0$ [as can easily be seen, this also follows
from the empirical equation (5.9)].

Furthermore it is evident from the H–T diagram that at temperatures below T_c the superconductor can exist both in the superconducting state [below the $H_c(T)$ curve] and in the normal state
[above the $H_c(T)$ curve], while at temperatures above T_c it can
only exist in the normal state.

5.2. Thermodynamics of the Transition from the Superconducting State to the Normal State

The transition of a superconductor from the superconducting
to the normal state is reversible; as shown by detailed experimental

studies this transition does not involve any irreversible expenditure of energy. Therefore the superconducting and normal states of a superconductor can, from the thermodynamic point of view, be regarded as two phases, and it is evident that in the H–T diagram the coexistence line of the two phases is the curve $H_c(T)$, which is the boundary of the region where each of the phases exists. This curve is quite similar to a saturation line for a phase transition, for example, the liquid–gas phase transformation in the p–T diagram.

In order to elucidate the features of the phase transition in a superconductor we seek an equation for the transition curve $H_c(T)$ using ordinary thermodynamic methods (similar to those used to derive the Clausius–Clapeyron equation).

We consider two phases – the superconducting phase (subscript s) and the normal phase (subscript n), which are in equilibrium at a temperature T, pressure p, and external magnetic field strength H_c.

As shown in Section 2.4 [Eq. (2.141)], the chemical potentials φ^* of the two phases are equal for two phases in equilibrium; in the case of interest

$$\varphi^*_s(p, T, H_c) = \varphi^*_n(p, T, H_c). \tag{5.10}$$

We change the temperature of each phase by dT and the magnetic field strength by dH_c (since we are interested in the form of the relation between H_c and T, the pressure in both phases will be regarded as constant). If we change the temperature and the field strength in such a way that the phases continue to remain in equilibrium at the new temperature $T + dT$, then it is clear that the potentials of the two phases will also be equal under these new conditions:

$$\varphi^*_s(p, T+dT, H_c+dH_c) = \varphi^*_n(p, T+dT, H_c+dH_c). \tag{5.11}$$

The function $\varphi^*(p, T + dT, H_c + dH_c)$ can be expanded in the series

$$\varphi^*(p, T+dT, H_c+dH_c) = \varphi^*(p, T, H_c) + \left(\frac{\partial \varphi^*}{\partial T}\right)_{H, p} dT +$$

$$+ \left(\frac{\partial \varphi^*}{\partial H_c}\right)_{T, p} dH_c + \cdots \tag{5.12}$$

Furthermore from Eq. (3.26) we find that

$$\varphi^* = u - Ts + pv - Hj$$

and, using (3.22),

$$Tds = du + p\,dv - H\,dj;$$

it follows that

$$d\varphi^* = -s\,dT + v\,dp - j\,dH. \tag{5.13}$$

From this it is evident that

$$\left(\frac{\partial \varphi^*}{\partial T}\right)_{H,\ p} = -s, \tag{5.14}$$

$$\left(\frac{\partial \varphi^*}{\partial H}\right)_{T,\ p} = -j, \tag{5.15}$$

$$\left(\frac{\partial \varphi^*}{\partial p}\right)_{T,\ H} = v. \tag{5.16}$$

Using Eqs. (5.14) and (5.15) we can write the series (5.12) in the following form (retaining only the first three terms):

$$\varphi^*(p,\ T + dT,\ H_c + dH_c) = \varphi^*(p,\ T,\ H_c) - sdT - jdH_c. \tag{5.17}$$

Substituting this expression into the left- and right-hand sides of Eq. (5.11), we find

$$\varphi^*_s(p,\ T,\ H_c) - s_s dT - j_s dH_c = \varphi^*_n(p,\ T,\ H_c) - s_n dT - j_n dH_c. \tag{5.18}$$

Since

$$\varphi^*_s(p,\ T,\ H_c) = \varphi^*_n(p,\ T,\ H_c),$$

we find that

$$\frac{dH_c}{dT} = \frac{s_n - s_s}{j_s - j_n} \tag{5.19}$$

(note that this equation was introduced for the case p = const).

Using Eqs. (5.4) and (5.7), Eq. (5.19) can be represented as

$$\frac{dH_c}{dT} = -\frac{4\pi(s_n - s_s)}{H_c v_s}.$$ (5.20)

This equation uniquely relates the slope of the transition line in the H–T diagram for the phase transition in a superconductor to the magnitude of the entropy differences for the corresponding phases and the critical magnetic field H_c at the given temperature.

We recall that the coexisting phases have identical temperature T and pressure p. When (5.20) is applied to systems involving a magnetic field, in agreement with Eq. (2.122) we add the condition of equal magnetic field strength H to these conditions:

$$H_1 = H_2.$$ (5.21)

Consequently the transformation of a superconductor from the superconducting to the normal state occurs at T = const, p = const, and H = const. The combined equation for the first and second laws of thermodynamics for a system in a magnetic field written in the form (3.25) for the weight-specific quantities

$$T ds = di^* - v\, dp + j\, dH$$

when applied to the present case assumes the form

$$T ds = di^*.$$ (5.22)

If we integrate this equation between arbitrary points 1 and 2 and assume that T remains constant during the process, we find

$$T(s_2 - s_1) = i^*_2 - i^*_1.$$ (5.23)

When applied to a phase transition from the superconducting to the normal state, the quantity $(i^*_2 - i^*_1)$ is the difference between the enthalpies of the coexisting phases, $(i^*_n - i^*_s)$. This enthalpy difference represents the heat of the phase transition in a superconductor. Denoting it by

$$q = i^*_n - i^*_s,$$ (5.24)

we find from (5.23) that

$$s_n - s_s = \frac{q}{T}.$$ (5.25)

In view of this equation, we find from (5.20) that

$$\frac{dH_c}{dT} = -\frac{4\pi q}{H_c T v_s}.$$

(5.26)

Since the heat q of the phase transition is positive, it follows from (5.26) that we always have

$$\frac{dH_c}{dT} < 0,$$

(5.27)

i.e., the critical magnetic field strength of a superconductor increases with decreasing temperature. This conclusion is in complete agreement with experimental data on the H_c—T relation (see Fig. 5.2., for example).

It follows from Eq. (5.26) that the heat of the phase transition is

$$q = -\frac{T H_c v_s}{4\pi} \frac{dH_c}{dT}.$$

(5.28)

It is evident from this that with $H_c = 0$ (i.e., $T = T_c$) the heat of the phase transformation goes to zero (as shown experimentally, dH_c/dT retains a finite value at any temperature). Thus along the transition curve at all points with $T < T_c$ there is an ordinary second-order phase transformation. We recall that the distinguishing sign of a first-order phase transformation is a discontinuous change in the first derivatives of the thermodynamic potentials at the transition point (i.e., i, j, and s in the present case). For a superconductor with $T < T_c$ the enthalpy increases discontinuously at the transition point by an amount

$$i^*{}_n - i^*{}_s = -\frac{T H_c v_s}{4\pi} \frac{dH_c}{dT},$$

(5.29)

and the magnetization, as seen from (5.4) and (5.7), discontinuously increases by

$$j_n - j_s = \frac{H v_s}{4\pi},$$

(5.30)

while the entropy, as follows from (5.20), discontinuously increases by

$$s_n - s_s = -\frac{H_c v_s}{4\pi} \frac{dH_c}{dT}.$$

(5.31)

The point $T = T_c$ is a special point in the curve for the transition from the superconducting to the normal state. As we see from Eqs. (5.28)-(5.31), when $T = T_c$ (while $H_c = 0$ at this temperature),

$$q = i^*_n - i^*_s = 0, \quad j_n - j_s = 0, \quad s_n - s_s = 0 \qquad (5.32)$$

at this point the phase transition from the superconducting to the normal state has no jumps in the first derivatives of the thermodynamic potentials. However, as will be shown below, in this case there are jumps in the second derivatives of the potentials at the transition point. Phase transitions having this feature are called second-order phase transitions.

Thus at all points on the curve with $T < T_c$ the transformation from the superconducting to the normal state is a first-order phase transformation while at the end point ($T = T_c$, $H_c = 0$) of the curve it degenerates into a second-order phase transformation.

Equation (5.26) for the phase transition in a superconductor was first obtained by W. Keesom in 1924 and is similar to the Clausius–Clapeyron equation for "ordinary" systems. The temperature T_c, (at $H_c = 0$) plays a role similar to that of the critical temperature in the liquid–vapor system (the heat of transformation, the jump in the entroy, etc., go to zero). However, at the critical point the liquid–vapor system does not have a second-order phase transformation (according to Ehrenfest's classification). In particular it should be noted that at the critical point a number of the second derivatives of the thermodynamic potentials such as the heat capacity c_p, $(\partial v/\partial T)_p$, and $(\partial v/\partial p)_T$ go to infinity.

We now consider the leftmost point on the transition curve in the H–T diagram – the point at $T = 0°K$. As noted previously, at this point $dH_c/dT = 0$. Consequently as we see from (5.28) and (5.31), in this case

$$q = i^*_n - i^*_s = 0 \quad \text{and} \quad s_n - s_s = 0. \qquad (5.33)$$

This conclusion is not so astonishing – the fact that $(s_n - s_s)$ goes to zero at $0°K$ is in complete agreement with the Nernst law. As for the magnetization, at this point, as at other points on the transition curve with $T < T_c$, it changes discontinuously, i.e.,

$$j_n - j_s \neq 0. \tag{5.34}$$

Since the entropy difference between the normal and super-conducting phases $(s_n - s_s)$ goes to zero at $T = T_c$ and at $0°K$, it is evident that the relation $(s_n - s_s) = f(T)$ passes through a maximum. This relation is shown in Fig. 5.4 for certain superconductors.

Equation (5.26) was derived assuming that the pressure of the external medium remains constant. It is of interest to explain the features of the phase transition in superconductors under conditions where the temperature of the superconductor remains constant while the pressure p and the strength H of the magnetic field change. Using Eq. (5.10) and assuming that the temperature of the phases remains constant, we let the pressures in each of the phases vary by an amount dp while the magnetic field strength varies by dH_c and use a method like that in the derivation of Eq. (5.26). We use the fact that in agreement with (5.16) we have

$$\left(\frac{\partial \varphi^*}{\partial p}\right)_{T, H_c} = v,$$

and find

$$\frac{dH_c}{dp} = \frac{v_s - v_n}{j_s - j_n}. \tag{5.35}$$

This equation shows how H_c varies with a change in the pressure of the medium (with the superconductor at a constant temperature).

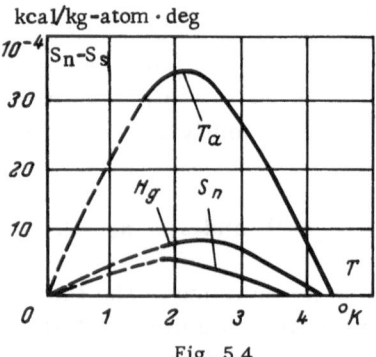

Fig. 5.4

Here v_n and v_s are the specific volumes of the superconductor in the normal and superconducting states, respectively.

In view of (5.4) and (5.7) we find

$$\frac{dH_c}{dp} = \frac{4\pi}{H_c} \frac{v_n - v_s}{v_s}. \tag{5.36}$$

Finally, knowing dH_c/dT with p = const (5.26), we can use the obvious relation

$$\left(\frac{\partial p}{\partial T}\right)_{H_c} = -\left(\frac{\partial T}{\partial H_c}\right)_p \left(\frac{\partial H_c}{\partial p}\right)_T \tag{5.37}$$

to show that

$$\left(\frac{\partial p}{\partial T}\right)_{H_c} = \frac{q}{(v_n - v_s)T}. \tag{5.38}$$

This equation is similar in form to the ordinary Clausius–Clapeyron equation and shows how the pressure should increase in a superconductor as its temperature changes such that the constant external magnetic field preserves the critical value.

The values of dH_c/dp and $(\partial p/\partial T)_{H_c}$ are quite perceptible. For example, it has been experimentally established for tin that as the pressure increases from 1 to 1750 kg/cm^2, the transition temperature falls by 0.1°K while H_c decreases by 14 Oe (i.e., for tin we have $(\partial T/\partial p)_{H_c} = -5.7 \times 10^{-5}$ deg·cm^2/kg and $dH_c/dp = 0.8 \times 10^{-2}$ Oe·cm^2/kg).

5.3. The Phase Diagram of a Superconductor

In Section 5.1 we considered one of the basic phase diagrams of a superconductor, the H–T diagram. We now turn to an analysis of other phase diagrams for a superconductor.

In the diagrams which we consider below we use not the weight-specific magnetization j but the volume-specific magnetization J. This is because the diagrams for J are simpler than those for j. Indeed, as we see from Eq. (5.3), the volume-specific magnetization J_s of a superconductor in the superconducting state is independent of temperature since

$$\varkappa_s = -\frac{1}{4\pi}; \tag{5.39}$$

consequently in these diagrams the isotherms coincide in the superconducting state.

As for the weight-specific magnetization j_s of the superconductor in the superconducting state, as seen from Eq. (5.4) it in general depends on the temperature since this equation contains the specific volume v_s of the superconductor, which varies with temperature Consequently, rigorously speaking, in diagrams for j the isotherms do not coincide with each other in the superconducting state.

In fact, this difference is of little importance in the precision of practical calculations using the diagrams.

We recall in this connection that the transformation from J to j is done using Eq. (3.9).

We consider the J−H diagram (this diagram has the same significance for a superconductor as the v−p diagram for "ordinary" systems).

In agreement with Eqs. (5.3) and (5.7) it is evident in the first place that $J_s < 0$ and decreases linearly with increasing H, while at $H = H_c$ it jumps discontinuously to $J_n = 0$ (Fig. 5.5). Here the magnitude of the jump (J_s-J_n) is

$$J_s - J_n = -\frac{H_c}{4\pi}. \tag{5.40}$$

since H_c decreases with increasing temperature in agreement with 5.9) (goes to zero at $T = T_c$), it is clear that the magnitude of the jump (j_s-J_n) also decreases with increasing temperature, going to zero at $T = T_c$.

In the second place it is clear that since J_s and J_n do not depend on the temperature, all of the isotherms for $T < T_c$ will coincide with each other in a J−H diagram, with the only difference being that the phase transition point (i.e., the point on the isotherm at which there is a jump J_s-J_n) will be different for different temperatures:

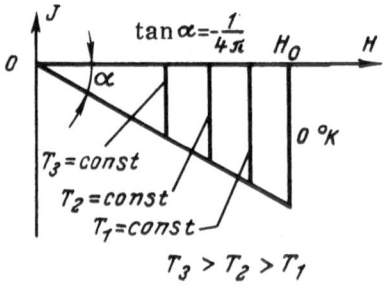

Fig. 5.5

the lower the temperature the lower H_c and farther left the phase transition point on the isotherm in the J–H diagram (Fig. 5.5). Here it is clear that in this diagram the isotherm T_c = const will coincide with the abscissa (so that with $T = T_c$ the transition occurs with $H = 0$, i.e., at the origin of coordinates in this diagram).

Figure 5.6 shows the B–H diagram of a superconductor. In view of (5.2) and (5.8) it is clear that in the superconducting state the isotherms in this diagram coincide with the abscissa and in the normal state they lie at an angle of 45° to the abscissa.

The T–S diagram of a superconductor is shown in Fig. 5.7. The shaded region between curves OBA and OCA represents the projection of the phase transition line from the superconducting state to the normal state (i.e., it is a two-phase region).

The boundary curve OBA corresponds to the superconducting phase while the boundary curve OCA corresponds to the normal phase. The length of the horizontal segment between the bounding curves OBA and OCA equals the entropy difference between the normal and superconducting phases, which goes to zero at 0°K and at T_c so that the bounding curves come together at these temperatures.

In agreement with the preceding discussion it is clear that since there is no temperature or magnetic field difference between phases when the superconducting and normal phases are in equilibrium, within the shaded region the lines H = const and the isotherms coincide in direction (horizontal). We shall now explore how the lines H = const run in the superconducting and normal regions.

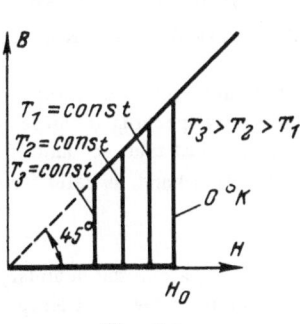

Fig. 5.6

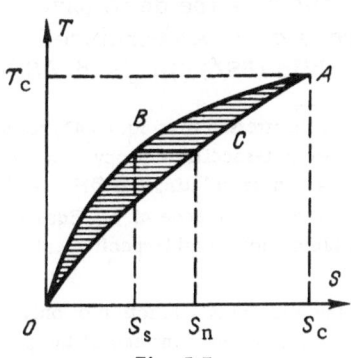

Fig. 5.7

In agreement with the Maxwell equation (3.31)

$$\left(\frac{\partial S}{\partial H}\right)_{T,\,p} = \left(\frac{\partial J}{\partial T}\right)_{H,\,p}.$$

Since in agreement with (5.3) J_s is independent of the temperature in the region of the superconducting state, we have from (3.31)

$$\left(\frac{\partial S_s}{\partial H}\right)_{T,\,p} = 0. \tag{5.41}$$

In the normal state region, in agreement with (5.7), we have $J_n = 0$ and we also find

$$\left(\frac{\partial S_n}{\partial H}\right)_{T,\,p} = 0 \tag{5.42}$$

from Eq. (3.31).

Equations (5.41) and (5.42) show that in both the superconducting and normal states, with T = const and p = const, the volume–specific entropy of a superconductor does not depend on the magnitude of the external magnetic field H. Thus the volume–specific entropy of a superconductor is a unique function of temperature.[†]

Furthermore in agreement with (1.49) we can write

$$\frac{dS}{dT} = \left(\frac{\partial S}{\partial T}\right)_{H,\,p} + \left(\frac{\partial S}{\partial H}\right)_{T,\,p}\frac{dH_c}{dT}, \tag{5.43}$$

where dS/dT is the derivative of the volume–specific entropy with respec to temperature[‡] along the boundary curve (OBA or OCA) while $(\partial S/\partial T)_{H,\,p}$ and $(\partial S/\partial H)_{T,\,p}$ are derivatives taken at the

[†]It should be stressed that Eq. (5.41) refers only to the volume-specific entropy. As for the weight-specific entropy of a superconductor, strictly speaking in agreement with (5.4), in calculating $(\partial s/\partial H)_{T,p}$ from Eq. (3.31) we must take account of the temperature dependence of v_s. Equation (5.42), on the other hand, is valid for both the volume- and weight-specific entropies.

[‡]Since the entropy is a function of only one variable, T or H, along the boundary curve (with p = const) instead of the partial derivative we have the total temperature derivative of S.

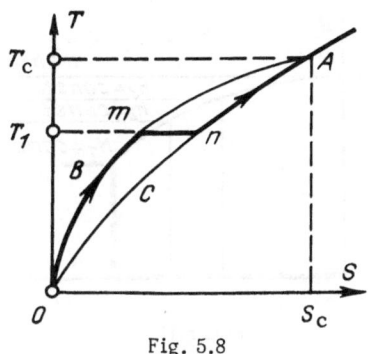

Fig. 5.8

intersection points of the corresponding lines H = const and the isotherms with this boundary curve.

In view of (5.41) and (5.42) we find that for both the super-conducting and normal phases we have

$$\frac{dS}{dT} = \left(\frac{\partial S}{\partial T}\right)_{H,\,p}. \tag{5.44}$$

It follows from this relation that the lines H = const in a T–S diagram coincide with the boundary curves OBA and OCA. Consider, for example, the line H_1 = const; with H = H_1 the transition from the superconducting to the normal state takes place at a certain temperature T_1. For temperatures T < T_1 the line H_1 = const coincides with the line OBA. At T = T_1 there is a phase transition and the line H = const coincides with the isotherm T_1 = const (section m–n). With a further increase in the temperature the line H_1 = const coincides with the line OCA (Fig. 5.8).

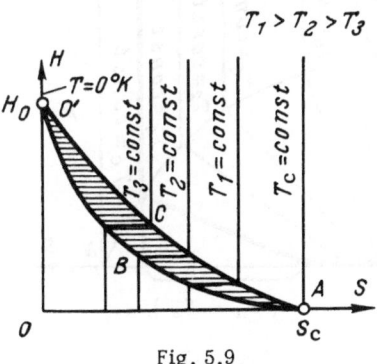

Fig. 5.9

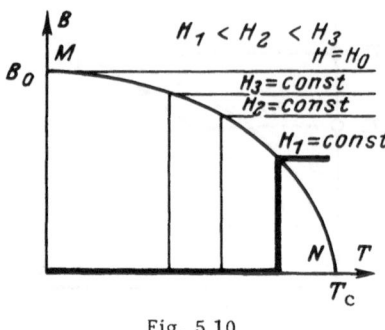

Fig. 5.10

We now consider the H–S diagram of a superconductor (Fig. 5.9), in which the region between the boundary curves of the phase transition from the superconducting to the normal state is shaded. The boundary curve O'BA corresponds to the superconducting phase while the boundary curve O'CA corresponds to the normal phase. Below this region lies the region of the superconducting state and above it lies the region of the normal state. The curves which bound the region of the phase transition meet along the ordinate axis at the point $H = H_0$ ($T = 0°K$) while along the abscissa (i.e., with $H_c = 0$) they meet at the entropy S_c corresponding to the temperature T_c.

The isotherms in this diagram are similar. Within the phase transition region the isotherms coincide with the lines $H = const$ and consequently in the H–S diagram in the phase transition region the isotherms are horizontal. As for the regions of the super-

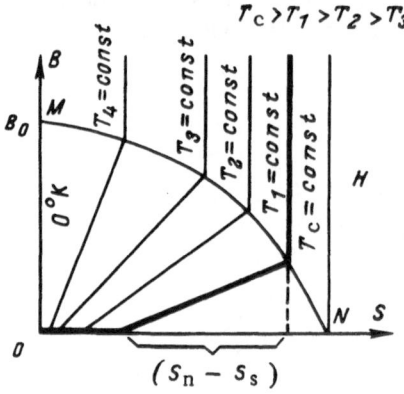

Fig. 5.11

conducting and normal states since the volume-specific entropies
of the superconducting and normal phases depend only on the temperature and do not change with H, in agreement with Eqs. (5.41)
and (5.42), the isotherms are vertical lines in these regions of
this diagram.

Figures 5.10 and 5.11 show the B–T and B–S diagrams for a
superconductor. The basic features of these diagrams are defined
by the fact that, in agreement with (5.2), the magnetic induction
field B in a superconductor is zero in the superconducting state
and in agreement with (5.8) coincides with H in the normal state.
It is clear from this that in a B–T diagram the whole region of
the superconducting state coincides with the abscissa. In Fig. 5.10
the line MN is the boundary curve which separates the two-phase
region from the normal-state region (the normal-state region lies
above the curve MN in this diagram). Since within the phase transition region a line H = const coincides with an isotherm, in this
diagram the lines H = const beneath the curve MN run vertically
while in the normal-state region these lines are horizontal (since
B = H in this region); the nature of the behavior of a line H = const
is shown in Fig. 5.10 using as an example the line H_1 = const illustrated with a heavy line.

In the B–S diagram (Fig. 5.11) the region of the superconducting
state (B = 0) merges with the abscissa. In the normal-state region
(beneath the curve MN) the isotherms are vertical (the entropy is
independent of H and consequently also of B). Within the two-phase
region beneath the boundary curve MN, the isotherms, as can easily
be shown, are lines whose slopes decrease with increasing temperature. The length of the segment along the abscissa under the part
of the isotherm which is in the two-phase region equals the entropy
jump during the phase transformation, $(S_n - S_s)$. The behavior of an
isotherm in this diagram is shown, for example, by the isotherm
T_1 = const illustrated in Fig. 5.11 by a heavy line.

5.4. The Heat Capacity in the Superconducting and Normal Phases – Rutgers Formula

In agreement with (3.56) the specific heat capacity $c_{H,p}$
of any magnet is defined as

$$c_{H,\,p} = T \left(\frac{\partial S}{\partial T} \right)_{H,\,p}.$$

We calculate the difference in the heat capacities $c_{H,p}$ of a super-conductor in the superconducting and normal phases at the phase transition temperature. Evidently (see Fig. 5.7) everywhere within the region of the phase transition OBAC the heat capacity $c_{H,p}$ is infinitely large since a line H = const runs horizontally in this region of the $T-S$ diagram so that we have $(\partial s/\partial T)_{H,p} = \infty$ on these parts of the lines H = const. Therefore when we speak of calculating the difference in the heat capacities of the coexisting supercon-ducting and normal phases at the transition line, we mean the heat capacities $c_{H,p}$ along the boundary curves on their one-phase sides. We denote these heat capacities in the superconducting and normal coexisting phases at the boundary curves by $c_{H,p}^s$ and $c_{H,p}^n$.

To calculate the difference between $c_{H,p}^n$ and $c_{H,p}^s$ one should differentiate Eq. (5.31) for the difference in the specific entropies of the coexisting phases with respect to temperature

$$s_n - s_s = -\frac{H_c v_s}{4\pi}\frac{dH_c}{dT}$$

Taking the total derivative with respect to temperature along the boundary curves, we find

$$\frac{ds_n}{dT} - \frac{ds_s}{dT} = -\frac{v_s}{4\pi}\left[H_c\frac{d^2H_c}{dT^2} + \left(\frac{dH_c}{dT}\right)^2 + \frac{H_c}{v_s}\frac{dv_s}{dT}\frac{dH_c}{dT}\right], \tag{5.45}$$

where dv_s/dT is the total derivative of the specific volume of the superconductor in the superconducting phase at the boundary curve OCA, taken along this curve.

It is also clear that using (1.49) we can write

$$\frac{ds}{dT} = \left(\frac{\partial s}{\partial T}\right)_{H,p} + \left(\frac{\partial s}{\partial H}\right)_{T,p}\frac{dH_c}{dT}, \tag{5.46}$$

which is a relation similar to (5.43) written for the volume-specific entropy while the other relation was for the weight-specific entropy.

Using (3.56) and (3.31) this relation can be put in the form

$$\frac{ds}{dT} = \frac{c_{H,p}}{T} + \left(\frac{\partial j}{\partial T}\right)_{H,p}\frac{dH_c}{dT}, \tag{5.47}$$

so that we can write

$$\frac{ds_{\mathrm{n}}}{dT} = \frac{c_{H,\,p}^{\mathrm{n}}}{T} + \left(\frac{\partial j_{\mathrm{n}}}{\partial T}\right)_{H,\,p} \frac{dH_{\mathrm{c}}}{dT} \tag{5.48}$$

and

$$\frac{ds_{\mathrm{s}}}{dT} = \frac{c_{H,\,p}^{\mathrm{s}}}{T} + \left(\frac{\partial j_{\mathrm{s}}}{\partial T}\right)_{H,\,p} \frac{dH_{\mathrm{c}}}{dT}, \tag{5.49}$$

where $(\partial j/\partial T)_{H,\,p}$ is the derivative of j, taken at the transformation point.

It is clear from (5.4) that[†]

$$\left(\frac{\partial j_{\mathrm{s}}}{\partial T}\right)_{H,\,p} = -\frac{H}{4\pi}\left(\frac{\partial v_{\mathrm{s}}}{\partial T}\right)_{H,\,p}, \tag{5.50}$$

while we find from (5.7) that

$$\left(\frac{\partial j_{\mathrm{n}}}{\partial T}\right)_{H,\,p} = 0. \tag{5.51}$$

Using these relations we find from (5.48) and (5.49) that

$$\frac{ds_{\mathrm{n}}}{dT} = \frac{c_{H,\,p}^{\mathrm{n}}}{T} \tag{5.52}$$

and

$$\frac{ds_{\mathrm{s}}}{dT} = \frac{c_{H,\,p}^{\mathrm{s}}}{T} - \frac{H_{\mathrm{c}}}{4\pi}\left(\frac{\partial v_{\mathrm{s}}}{\partial T}\right)_{H,\,p} \frac{dH_{\mathrm{c}}}{dT}. \tag{5.53}$$

Furthermore, in agreement with (1.49) we write

$$\frac{dv_{\mathrm{s}}}{dT} = \left(\frac{\partial v_{\mathrm{s}}}{\partial T}\right)_{H,\,p} + \left(\frac{\partial v_{\mathrm{s}}}{\partial H}\right)_{T,\,p} \frac{dH_{\mathrm{c}}}{dT}, \tag{5.54}$$

it being understood that the partial derivatives which enter this formula are taken at the transition point.

In view of (5.52)–(5.54) we find from (5.45) that

$$c_{H,\,p}^{\mathrm{s}} - c_{H,\,p}^{\mathrm{n}} = \frac{v_{\mathrm{s}}T}{4\pi}\left\{H_{\mathrm{c}}\frac{d^2 H_{\mathrm{c}}}{dT^2} + \left(\frac{dH_{\mathrm{c}}}{dT}\right)^2 + \right.$$

$$\left. + \frac{H_{\mathrm{c}}}{v_{\mathrm{s}}}\frac{dH_{\mathrm{c}}}{dT}\left[2\left(\frac{\partial v_{\mathrm{s}}}{\partial T}\right)_{H,\,p} + \left(\frac{\partial v_{\mathrm{s}}}{\partial H}\right)_{T,\,p}\frac{dH_{\mathrm{c}}}{dT}\right]\right\}. \tag{5.55}$$

[†]We call attention to the fact that Eq. (5.50) is written for the weight-specific magnetization j_{s}. As for the volume-specific magnetization J_{s}, as we see from (5.3), $(\partial J_{\mathrm{s}}/\partial T)_{H,\,p} = 0$.

Since for a superconductor we can assume with good precision that $(\partial v_s/\partial T)_{H,\,p} \approx 0$ at the temperature of the transition from the super-conducting to the normal state,[†] and since the magnetostriction of a superconductor can be regarded as practically zero in the super-conducting state, i.e., $(\partial v_s/\partial H)_{T,\,p} = 0$, Eq. (5.55) can be written as follows:

$$c_{H,\,p}^{s} - c_{H,\,p}^{n} = \frac{v_s T}{4\pi}\left[H_c \frac{d^2 H_c}{dT^2} + \left(\frac{dH_c}{dT}\right)^2 \right]. \tag{5.56}$$

Equation (5.56) can be obtained in a different way. Replacing the weight-specific entropy by the volume-specific value S in Eq. (5.31) and using the obvious relation

$$s = Sv, \tag{5.57}$$

we find

$$S_n \frac{v_n}{v_s} - S_s = -\frac{H_c}{4\pi}\frac{dH_c}{dT}; \tag{5.58}$$

where v_n and v_s are the specific volumes of the superconductor in the normal and superconducting phases at the transition point.

The change in the volume of the superconductor during the phase transformation is so small that with great accuracy we can assume that

$$\frac{v_n}{v_s} \approx 1. \tag{5.59}$$

Using this relation, Eq. (5.58) can be put in the form

$$S_n - S_s = -\frac{H_c}{4\pi}\frac{dH_c}{dT}, \tag{5.60}$$

whence

$$\frac{dS_c}{dT} - \frac{dS_s}{dT} = -\frac{1}{4\pi}\left[H_c \frac{d^2 H_c}{dT^2} + \left(\frac{dH_c}{dT}\right)^2 \right]. \tag{5.61}$$

Using the previously found relation for the volume-specific entropy, Eq. (5.44), we find from (5.61) that

$$C_{H,p}^{s} - C_{H,p}^{n} = \frac{T}{4\pi}\left[H_c \frac{d^2 H_c}{dT^2} + \left(\frac{dH_c}{dT}\right)^2 \right]. \tag{5.62}$$

Finally, since

$$c_{H,\,p} = C_{H,\,p}v, \tag{5.63}$$

[†] We recall in this regard that in agreement with the Nernst law the derivative $(\partial v/\partial T)_p$ decreases in approaching absolute zero, such that

$$\lim_{T \to 0\ °K} \left(\frac{\partial v}{\partial T}\right)_p = 0$$

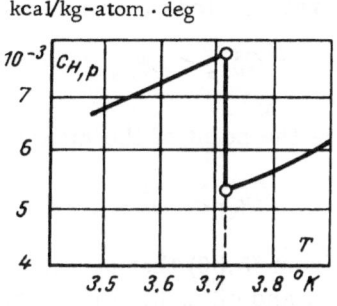

Fig. 5.12

by transforming from volume- to weight-specific heat capacities in (5.62) we find using (5.59)

$$c_{H,p}^{s} - c_{H,p}^{n} = \frac{v_s T}{4\pi} \left[H_c \frac{d^2 H_c}{dT^2} + \left(\frac{dH_c}{dT} \right)^2 \right],$$

which agrees with (5.56).

From Eq. (5.56) we find that when $H_c = 0$ (i.e., $T = T_c$)

$$c_{H,p}^{s} - c_{H,p}^{n} = \frac{v_s T}{4\pi} \left(\frac{dH_c}{dT} \right)^2. \tag{5.64}$$

This equation, which determines the size of the jump in the heat capacity $c_{H,p}$ at the phase transformation from the superconducting state to the normal state in the absence of an external magnetic field, is called the Rutgers formula.

It is obvious from the Rutgers formula that since the right-hand side of Eq. (5.64) is always positive, with $H_c = 0$ we always have $c_{H,p}^{s} > c_{H,p}^{n}$, i.e., the heat capacity $c_{H,p}$ decreases discontinuously during the transformation of a superconductor from the superconducting to the normal state. Figure 5.12 shows experimental data on the heat capacity of tin; this graph clearly shows a jump in the heat capacity at the transformation point from the superconducting to the normal state.

As to the sign of the difference $(c_{H,p}^{s} - c_{H,p}^{n})$ with $T < T_c$, as we see from what was said above,

$$C_{H,p}^{s} - C_{H,p}^{n} = - T \frac{d}{dT} (S_n - S_s), \tag{5.65}$$

while the relation $(S_n - S_s) = f(T)$ passes through a maximum

(Fig. 5.4) and at temperatures less than the maximum we have

$$C_{H,\,p}^{s} < C_{H,\,p}^{n}, \tag{5.66}$$

while everywhere above the point of the maximum we have

$$C_{H,\,p}^{s} > C_{H,\,p}^{n}. \tag{5.67}$$

Finally, we see from (5.56) that at 0°K the difference between the heat capacities $c_{H,\,p}^{s}$ and $c_{H,\,p}^{n}$ is zero. This conclusion is obvious since the heat capacity of matter in the condensed state at 0°K is zero in agreement with the Nernst law.

5.5. Magnetostriction of a Superconductor

In a superconductor, as in all magnets, there is a magnetostriction effect. The change in the volume of the superconductor with a change in the external magnetic field strength is described by Eqs. (3.140) and (3.141).

In a superconductor in the normal state $\chi_n = 0$ and consequently there is no magnetostriction. In the superconducting state, in agreement with (5.5),

$$\chi_s = -\frac{v_s}{4\pi},$$

and from Eqs. (3.140) and (3.141) we find for isothermal magnetostriction

$$v_s\,(T,\,p,\,H) - v_s\,(T,\,p,\,H=0) = \left(\frac{\partial v_s}{\partial p}\right)_{T,\,H} \frac{H^2}{8\pi} \tag{5.68}$$

and for adiabatic magnetostriction

$$v_s(s,\,p,\,H) - v_s\,(s,\,p,\,H=0) = \left(\frac{\partial v_s}{\partial p}\right)_{s,H} \frac{H^2}{8\pi}. \tag{5.69}$$

On the other hand, a change in the volume of the superconductor can be formally described as a change in v due to a certain additional pressure Δp:

$$v_s(p + \Delta p,\,H=0) - v_s(p,\,H=0) = \left(\frac{\partial v_s}{\partial p}\right)_H \Delta p. \tag{5.70}$$

We can see from a comparison of (5.68) and (5.69) with (5.70) that

$$\Delta p = \frac{H^2}{8\pi} \qquad (5.71)$$

and consequently

$$v_s(p,\ H) - v_s(p,\ H=0) = v_s\left(p + \frac{H^2}{8\pi},\ H=0\right) - v_s(p,\ H=0). \qquad (5.72)$$

Furthermore, since we always have $(\partial v/\partial p)_T < 0$ and $(\partial v/\partial p)_s < 0$, it is clear from Eqs. (5.68) and (5.69) that we always have

$$v_s(p,\ H) < v_s(p,\ H=0), \qquad (5.73)$$

i.e., the volume of a superconductor in the superconducting state decreases with increasing strength of the external magnetic field.

As for the magnetoelastic effect — the change in magnetization with a change in the external pressure — the magnitude of this effect is determined as follows for a superconductor in the superconducting state. For a pressure change under isothermal conditions we have

$$j(T,\ p + \Delta p,\ H) - j(T,\ p,\ H) = \int_p^{p+\Delta p} \left(\frac{\partial j}{\partial p}\right)_{T,\ H} dp, \qquad (5.74)$$

while for an adiabatic pressure change

$$j(s,\ p + \Delta p,\ H) - j(s,\ p,\ H) = \int_p^{p+\Delta p} \left(\frac{\partial j}{\partial p}\right)_{s,\ H} dp. \qquad (5.75)$$

Furthermore we find from (3.10), using (5.5), that

$$\left(\frac{\partial j}{\partial p}\right)_{T,\ H} = -\frac{H}{4\pi}\left(\frac{\partial v_s}{\partial p}\right)_{T,\ H} \qquad (5.76)$$

and

$$\left(\frac{\partial j}{\partial p}\right)_{s,\ H} = -\frac{H}{4\pi}\left(\frac{\partial v_s}{\partial p}\right)_{s,\ H}. \qquad (5.77)$$

Using these relations we find from (5.74) and (5.75) that

$$j(T, p+\Delta p, H) - j(T, p, H) = -\frac{H}{4\pi} \int_{p}^{p+\Delta p} \left(\frac{\partial v_s}{\partial p}\right)_{T, H} dp \qquad (5.78)$$

and

$$j(s, p+\Delta p, H) - j(s, p, H) = -\frac{H}{4\pi} \int_{p}^{p+\Delta p} \left(\frac{\partial v_s}{\partial p}\right)_{s, H} dp. \qquad (5.79)$$

Since $(\partial v_c/\partial p)_H$ changes little with a change of pressure, these equations can be put in the form

$$j(T, p+\Delta p, H) - j(T, p, H) = -\frac{H}{4\pi} \left(\frac{\partial v_s}{\partial p}\right)_{T, H} \Delta p \qquad (5.80)$$

and

$$j(s, p+\Delta p, H) - j(s, p, H) = -\frac{H}{4\pi} \left(\frac{\partial v_s}{\partial p}\right)_{s, H} \Delta p. \qquad (5.81)$$

Since $(\partial v_c/\partial p)_H < 0$ it follows that the magnetization of a superconductor increases with increasing pressure in the superconducting state.

CHAPTER 6

Surface Phenomena

6.1. Some Basic Properties of Surfaces Separating Phases

It will be recalled from general physics that there is a so-called surface tension at the interface between two phases. This force tends to minimize the surface area of a liquid.

What exactly is surface tension? To answer this question we consider a scheme for the surface layer of a liquid at its boundar with the vapor (Fig. 6.1).

It is well known that the intermolecular interaction forces in a liquid are greater than in the saturated vapor of this liquid. Consider a molecule located in the bulk of the liquid (A in Fig. 6.1).[†] Obviously this molecule is acted upon by attractive forces toward other molecules directly touching it. Since these molecules surround molecule A on all sides, naturally the resultant effect of all of the intermolecular forces acting on molecule A is zero. Molecule B, which lies at the surface separating the phases, sees completely different conditions. Molecule B is acted upon by molecular attraction toward the molecules in the liquid which are beside and below it, while above molecule B the attractive forces arise from molecules in the vapor phase above the surface of the liquid. It

---

[†]The discussion given here is quite schematic and is not rigorous; one should consider not individual molecules but molecular clusters, etc. However the discussion does lead to qualitatively true results.

Fig. 6.1

should be noted that since the density of the vapor is much less
than the density of the liquid (far from the critical point, where
these densities are equal), the distance between molecules in the
vapor phase will be substantially greater than that between molecules
in the liquid phase. Clearly then the attractive forces between
molecule B and the molecules in the vapor phase are substantially
less than the attractive forces between molecule B and the molecules
in the liquid phase. Consequently the total effect of the intermolecular forces acting on molecule B is not zero. We see that this force
is directed into the liquid along the normal to the liquid surface.
Since these considerations apply to all the molecules located at the
surface of the liquid, the surface layer exerts a pressure on the
whole volume of liquid. This pressure is called the internal pressure. The internal pressure of a liquid is quite large. Calculations show that, for example, for water at atmospheric pressure
and at 100°C, the internal pressure is about 5800 kg/cm^2. It should
be stressed that the internal pressure cannot be measured directly;
if we place apparatus in the liquid then a surface layer of liquid
is formed at the boundary with the apparatus which "isolates" it
from the internal pressure.

The pressure in the surface layer of a liquid arises not
only from the outer layer of molecules but also from the layer
lying directly below it. Indeed, for example, for molecule C located
in the second layer (from the liquid surface) attractive forces from
many layers of molecules act downward and sideward while only
one layer of the liquid and the molecules of the vapor phase exert
upward forces. It is clear from this that the sum of the intermolecular forces acting on molecule C is also not zero and consequently the second-layer molecules also exert pressure on the
inner layers of liquid. However, it should be stressed that the
force acting on molecule C is substantially less than that acting

on molecule B, because the intermolecular attractive force di-
minishes rapidly with distance and therefore a given molecule
interacts most strongly with those molecules which are located
near it; molecules located farther away interact more weakly
with a given molecule. Since molecule C is surrounded by one layer
of molecules on all sides including surface molecules (sphere of
radius r_1 in Fig. 6.1), it is clear that the overall forces acting on
molecule C will be determined by the attraction of molecules lying
inside the sphere r_1 and naturally will be less than the overall
forces acting on molecule B.

Similar reasoning also applies to the third, fourth, fifth, etc.,
monomolecular layers. The further from the surface the molecule,
the smaller the overall force acting on it; even at a distance of
several intermolecular layers from the surface the overall force
acting on the molecules becomes practically zero (Fig. 6.2).

The layer of liquid at the boundary with its vapor in which
the overall force acting on a molecule is nonzero is called the surface layer.

It should be stressed that the greater the density of the vapor
phase, i.e., the more molecules of the vapor phase are near the
surface of the liquid, the stronger the interaction between the
molecules of the liquid and gas phases, and consequently the lower
the overall force acting on a molecule in the surface layer. Thus
in approaching the critical point the magnitude of the force acting
on the liquid at its interface with the vapor decreases (i.e., as we
see below, the surface tension decreases for a liquid).

It should be understood that since a molecule located in the
surface layer is acted upon by forces directed into the liquid, if
any molecule is to be removed from the depth of the liquid to the
surface a certain amount of work must be done against these forces.

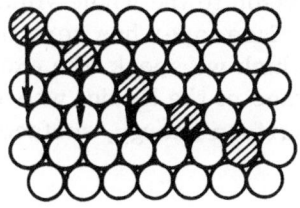

Fig. 6.2

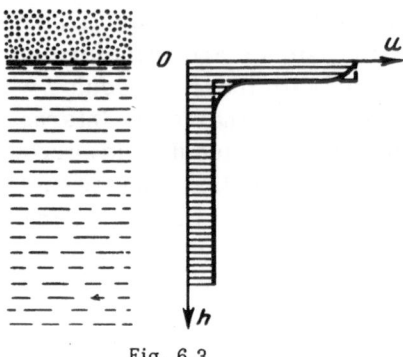

Fig. 6.3

It is clear from this that molecules in the surface layer have an additional energy in comparison to molecules deep within the liquid, equal to the work done in moving the molecule into the surface layer. Thus the surface layer has excess internal energy with respect to the remaining volume of liquid; this is called the surface energy.

It is clear in agreement with what was said above that the energies of individual monomolecular layers in the surface layer of a liquid are greater the closer the monomolecular layer is to the surface of the liquid (Fig. 6.3). Indeed, within the surface layer the internal energy of the liquid varies with height from the internal energy of the liquid deep within it to the value at the liquid surface in the outer molecular layer. However, since the surface layer is very thin (its thickness is of the order of several monomolecular layers), we can in practice regard the energy of the surface layer as constant over the whole thickness of the layer. Moreover, since the volume of the surface layer is usually† negligibly small in comparison to the whole volume of the liquid, we can assume that the surface layer has zero thickness and that the special properties of which we speak (the excess energy, etc.) appear only at the surface of the liquid, whose thickness is zero. Thus we speak of the surface energy, surface heat capacity, surface entropy, etc.

---

†With the exception of certain special materials (for example, microscopic liquid drops).

6.2. Surface Tension

The most important property of the surface layer is the surface tension. How does it reveal itself?

It is not hard to show that since the surface layer has excess energy, the surface of a liquid always tends to contract.

It was remarked above that the surface layer has additional (surface) energy. Since the energy is an extensive property it is clear that the surface energy is proportional to the amount of liquid surface — the greater the surface area the greater the number of molecules which must be extracted from the depths of the liquid to make up the surface and consequently the greater the work expended against internal pressure forces.

This question can also be examined from a different viewpoint. If the increase in surface area involves the performance of work, then there are forces in a plane tangent to the surface which tend to inhibit the increase in the surface area of the liquid. These forces evidently lie along the normal to the perimeter of the surface.

We consider the work expended in increasing the surface $\mathfrak{S}$ (Fig. 6.4) at a constant temperature T. We denote the perimeter of this arbitrary surface by l. We let the surface area $\mathfrak{S}$ increase in such a way that the perimeter l expands, moving an amount dx at all points. The force acting on a unit length of perimeter will be denoted by σ. Then the force acting on the whole perimeter will be σl, and the work against these forces which we want will be given by

$$dL_{\mathfrak{S}} = -\sigma l\, dx, \tag{6.1}$$

i.e.,

$$dL_{\mathfrak{S}} = -\sigma\, d\mathfrak{S}, \tag{6.2}$$

since it is clear that

$$l\, dx = d\mathfrak{S} \tag{6.3}$$

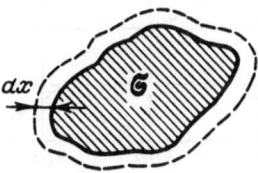

Fig. 6.4

TABLE 6.1

Material	$\sigma,$ ergs/cm^2	Material	$\sigma,$ ergs/cm^2
Ammonia	21.2	Ethyl alcohol	21.8
Acetone.	23.7		
Benzene.	29.0	Toluene	28.5
		Carbon dioxide.	1.2
Water . . . . : . .	72.75	Chlorobenzene.	33.5
Glycerin	59.4	Carbon tetra-	
		chloride	26.9
Methyl alcohol	22.6	Ethyl ether	17.0

The quantity σ is called the coefficient of surface tension (or simply the surface tension) and has the dimensions of force per unit length (or, what is the same, energy per unit area).

As an illustration Table 6.1 gives the surface tension coefficients for several materials at 20°C.

The surface tension for a given material is a unique function of temperature (this fact should be particularly stressed: since σ characterizes the properties of coexisting phases it depends only on one parameter — the temperature). With increasing temperature, i.e., as we approach the critical point, where the difference in the density of the liquid and its vapor is ever smaller, σ decreases, reaching zero at the critical point. The temperature dependence of the surface tension of water is shown in Fig. 6.5.

Thermodynamics does not allow us to relate σ to other thermodynamic properties and at present no rigorous theory of the temperature dependence of the surface tension exists. Therefore all of the existing formulas for the temperature dependence of the surface tension are empirical or semiempirical.

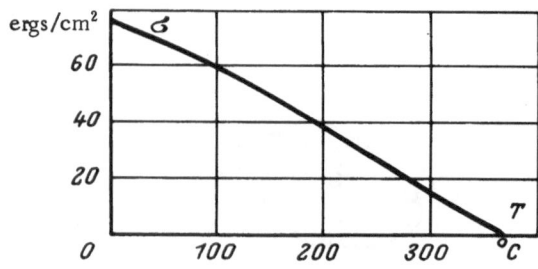

Fig. 6.5

6.3. Basic Thermodynamic Relations
for Surfaces

We consider a thermodynamic system composed of a surface having no thickness.

By comparing Eq. (6.2) for the work done on such a system

$$dL_{\mathfrak{S}} = -\sigma d\mathfrak{S}$$

with the general expression (1.25)

$$dL^* = \xi dX,$$

It is clear that in the system of interest the generalized coordinate is the surface area $\mathfrak{S}$ while the generalized force, taken with a minus sign, is the surface tension $(-\sigma)$.

The combined equation of the first and second laws of thermodynamics for such a system is, in agreement with (1.28a),

$$TdS = dU - \sigma d\mathfrak{S}. \tag{6.4}$$

Since the entropy, internal energy, etc., are state functions they naturally have the property of additivity. For the two-dimensional system of interest this means that

$$S = s_{\mathfrak{S}}\mathfrak{S}; \tag{6.5}$$

$$U = u_{\mathfrak{S}}\mathfrak{S} \tag{6.6}$$

etc. Here $s_{\mathfrak{S}}$ and $u_{\mathfrak{S}}$ are the area-specific values of S and U (since weight and volume are irrelevant for a two-dimensional system the specific values can only be referred to unit area).

On the other hand, since $s_{\mathfrak{S}}$ and $u_{\mathfrak{S}}$ depend only on the temperature, it is clear that

$$s_{\mathfrak{S}} = \left(\frac{\partial S}{\partial \mathfrak{S}}\right)_T, \tag{6.7}$$

$$u_{\mathfrak{S}} = \left(\frac{\partial U}{\partial \mathfrak{S}}\right)_T, \tag{6.8}$$

We can determine $s_{\mathfrak{S}}$ without difficulty using the Maxwell equation (2.150a), which, when applied to the present case, is written in the form

$$\left(\frac{\partial S}{\partial \mathfrak{S}}\right)_T = -\left(\frac{\partial \sigma}{\partial T}\right)_{\mathfrak{S}}. \tag{6.9}$$

Since σ, as noted above, depends only on the temperature, the partial derivative of σ with respect to T can be replaced by a total derivative, and we find, using (6.7),

$$s_{\mathfrak{S}} = -\frac{d\sigma}{dT}. \tag{6.10}$$

In this case Eq. (6.5) for the total entropy of the surface assumes the form

$$S = -\frac{d\sigma}{dT}\,\mathfrak{S}. \tag{6.11}$$

Since the surface tension always decreases with temperature, we have $d\sigma/dT < 0$ and consequently always $S > 0$.

We can determine $u_{\mathfrak{S}}$ as follows. We write Eq. (6.4) in the form

$$dU = TdS + \sigma d\mathfrak{S},$$

and find

$$\left(\frac{\partial U}{\partial \mathfrak{S}}\right)_T = T\left(\frac{\partial S}{\partial \mathfrak{S}}\right)_T + \sigma, \tag{6.12}$$

whence using (6.9) we find

$$u_{\mathfrak{S}} = \left(\frac{\partial U}{\partial \mathfrak{S}}\right)_T = \sigma - T\frac{d\sigma}{dT}. \tag{6.13}$$

Equation (6.6) assumes the following form for the total internal energy of the surface:

$$U = \left(\sigma - T\frac{d\sigma}{dT}\right)\mathfrak{S}. \tag{6.14}$$

Finally, the relation for the area-specific heat capacity defined by the obvious equation

$$c_{\mathfrak{S}} = T\left(\frac{\partial s_{\mathfrak{S}}}{\partial T}\right)_{\mathfrak{S}}, \tag{6.15}$$

taking (6.10) into account, is written in the form

$$c_{\mathfrak{S}} = -T\frac{d^2\sigma}{dT^2}. \qquad (6.16)$$

In order to get some idea of the magnitudes of $s_{\mathfrak{S}}$, $u_{\mathfrak{S}}$ and $c_{\mathfrak{S}}$, we give their numerical values for water at 100°C

$\sigma = 58.85$ ergs/cm^2 = 14.1×10^{-10} kcal/cm^2
$s_{\mathfrak{S}} = 0.197$ erg/cm$^2 \cdot$ deg = 0.047×10^{-10} kcal/cm$^2 \cdot$ deg
$u_{\mathfrak{S}} = 132.4$ ergs/cm^2 = 31.7×10^{-10} kcal/cm^2
$c_{\mathfrak{S}} = -1.49$ ergs/cm$^2 \cdot$ deg = -0.36×10^{-10} kcal/cm$^2 \cdot$ deg

We note yet another characteristic feature. We calculate the free energy of the surface; evidently in agreement with the general definition of free energy (2.12) the expression for the free energy of a surface is written in the form

$$F = U - TS.$$

Substituting the values of U and S from Eqs. (6.14) and (6.11) into this expression, we find

$$F = \sigma\mathfrak{S}. \qquad (6.17)$$

Furthermore, we can determine the amount of work done by the system of interest in changing the area in an isothermal process; using Eq. (6.2), integrating (6.2) and assuming that σ remains constant in an isothermal process, we find

$$L_{1-2} = -\sigma(\mathfrak{S}_2 - \mathfrak{S}_1), \qquad (6.18)$$

where the subscripts 1 and 2 refer, respectively, to the initial and final states of the system.

Evidently by using (6.17) we can put Eq. (6.18) in the following form:

$$L_{1-2} = F_1 - F_2. \qquad (6.19)$$

This is understandable since, as we know from general thermodynamics, the work done by a system in an isothermal process equals the change in its free energy.

We now consider the conditions for equilibrium in an isolated homogeneous† two-dimensional system. All of the discussion is similar to that given in Section 2.3 for the general case of an isolated homogeneous system.

For the isolated two-dimensional system of interest it is obvious that the conditions

$$U_{syst} = \text{const} \tag{6.20}$$

and

$$\mathfrak{S}_{syst} = \text{const} \tag{6.21}$$

are valid (the latter condition is equivalent to the condition $V_{syst} = \text{const}$ for ordinary three-dimensional systems).

We conceptually divide the system under study into two subsystems 1 and 2. Obviously

$$U_{syst} = U_1 + U_2 \tag{6.22}$$

and

$$\mathfrak{S}_{syst} = \mathfrak{S}_1 + \mathfrak{S}_2, \tag{6.23}$$

where the subscripts 1 and 2 refer to the first and second subsystems.

It follows from (6.22) and (6.23), using (6.20) and (6.21), that

$$dU_1 = -dU_2 \tag{6.24}$$

and

$$d\mathfrak{S}_1 = -d\mathfrak{S}_2. \tag{6.25}$$

Recall that, as shown in Chapter 2, the entropy of an isolated system in equilibrium retains a constant (maximum) value

$$dS_{syst} = 0. \tag{6.26}$$

---

†By homogeneous system we mean a system composed of a single material in a single phase.

Since the entropy is additive, we have

$$S_{\text{syst}} = S_1 + S_2, \tag{6.27}$$

whence, using (6.26), we find that

$$dS_1 + dS_2 = 0. \tag{6.28}$$

From Eq. (6.4) we find for the first subsystem

$$dS_1 = \frac{1}{T_1} dU_1 - \frac{\sigma_1}{T_1} d\mathfrak{S}_1 \tag{6.29}$$

while for the second subsystem

$$dS_2 = \frac{1}{T_2} dU_2 - \frac{\sigma_2}{T_2} d\mathfrak{S}_2. \tag{6.30}$$

Substituting (6.29) and (6.30) into (6.28) and using (6.24) and (6.25), we have

$$\left(\frac{1}{T_1} - \frac{1}{T_2}\right) dU_1 - \left(\frac{\sigma_1}{T_1} - \frac{\sigma_2}{T_2}\right) d\mathfrak{S}_1 = 0. \tag{6.31}$$

It is to be understood that the differentials dU_1 and $d\mathfrak{S}_1$ are mutually independent; in principle the internal energy of the subsystem and its surface area can vary independently of each other. In this case it is clear that if the left-hand side of Eq. (6.31) is to be zero the coefficients of the differentials dU_1 and $d\mathfrak{S}_1$, must be separately equal to zero, i.e.,

$$\frac{1}{T_1} - \frac{1}{T_2} = 0 \tag{6.32}$$

and

$$\frac{\sigma_1}{T_1} - \frac{\sigma_2}{T_2} = 0. \tag{6.33}$$

It follows from (6.32) that

$$T_1 = T_2, \tag{6.34}$$

while from (6.33), using Eq. (6.34), we find that

$$\sigma_1 = \sigma_2. \tag{6.35}$$

The results of this treatment do not change regardless what two subsystems we divide the system into.

Thus we arrive at the obvious conclusion that in an isolated two-dimensional thermodynamic system in equilibrium the temperature and the surface tension σ are the same in all parts of the system. Clearly, this conclusion is a particular case of the most general equilibrium conditions for an isolated homogeneous thermodynamic system (2.120)-(2.122) formulated in Chapter 2.

6.4. The Effect of Surface Phenomena on the Thermodynamic Properties of a System

If we now consider not a two-dimensional system but an ordinary three-dimensional system composed of a pure substance, the combined equation for the first and second laws of thermodynamics, taking surface effects into account in agreement with Eqs. (1.28a) and (6.2), is written as follows:

$$TdS = dU + pdV - \sigma d\mathfrak{S}. \qquad (6.36)$$

As for the thermodynamic properties of such systems, the relations for U, I, F, Φ, S, and C for a system are obtained using the additivity of these quantities, using Eqs. (6.11), (6.14), and (6.16),

$$U = uG + \left(\sigma - T\frac{d\sigma}{dT}\right)\mathfrak{S}, \qquad (6.37)$$

$$I = iG + \left(\sigma - T\frac{d\sigma}{dT}\right)\mathfrak{S}, \qquad (6.38)$$

$$F = fG + \sigma\mathfrak{S}, \qquad (6.39)$$

$$\Phi = \varphi G + \sigma\mathfrak{S}, \qquad (6.40)$$

$$S = sG - \frac{d\sigma}{dT}\mathfrak{S}, \qquad (6.41)$$

and

$$C = cG - T\frac{d^2\sigma}{dT^2}\mathfrak{S}, \qquad (6.42)$$

where G is the weight of the material in the system, f and φ are determined respectively by Eqs. (2.74) and (2.52), C is the total

heat capacity of the system, and c is the weight-specific heat capacity

It is clear that when the ratio of the surface area of a liquid to its volume is comparatively small the "contribution" of the surface layer of the liquid to the thermodynamic functions of the liquids is negligible. On the other hand, the role of the surface layer increases when the ratio of area to volume is large. This condition occurs in two cases: when we have small amounts of liquid (recall that the ratio of surface area to volume for a sphere is inversely proportional to the sphere radius) or when we somewhat increase the surface area of a comparatively large mass of liquid slightly (for example, by dividing the liquid by planar surfaces, creating a foam, etc.).

Table 6.2 shows a comparison of surface and volume parts of such thermodynamic quantities as the free energy, entropy, internal energy, and heat capacity, of water at 100°C in two cases for a sphere of 100 mm diameter (in this case the surface-to-volume ratio is 0.6 cm^{-1}) and for fine water drops of $0.1\,\mu$ diameter[†] (the ratio of surface to volume is 6×10^5 cm^{-1}). As we see from this table, for fine drops the surface and volume parts of the thermodynamic quantities become commensurable. Note also that if the amount of water in a sphere of 100 mm diameter is distributed over a horizontal surface (for example, over the surface of a liquid which does not mix with water) in a planar film 10^{-4} mm thick, then, for example, the free energy of the surface layer makes up about 2% of the overall free energy of the liquid.

Obviously the smallest thickness a surface layer can have is one molecule – a so-called monomolecular layer. A monomolecular layer is obtained when a liquid spreads over a solid surface or over a free surface of another liquid with which it does not mix (for example, kerosene on water).

We noted above that the surface tension force tends to reduce the surface area[‡]. Thus the surface film behaves like a flexible elastic film. In those cases where all of the other forces besides surface tension are negligibly small, the surface tension forces will evidently produce a shape such that the surface area is least, i.e., a sphere. Under the usual conditions for comparatively large amounts of liquid the surface tension forces are small in comparison to the weight of the liquid, and the surface of liquid in a vessel is horizontal. It is clear from this that surface tension will affect the shape of a liquid only in those cases where the effect of gravitation-

---

[†]For comparison, note that the size of a water vapor molecule found from data on the viscosity of water vapor amounts to 2×10^{-8} cm.

[‡]It is not hard to explain the tendency of a liquid to reduce its surface area. As shown in Section 2.1, for a system having a constant volume with T = const and V = const, the values of f, G, and σ are constant and F depends uniquely on the surface area $\mathfrak{S}$ such that the lower $\mathfrak{S}$, the lower F. Thus, the free energy of the system is least when the surface area of the liquid has the lowest possible value.

TABLE 6.2

Thermodynamic function	100-mm-diameter sphere	10^{-4}-mm-diameter drop
F, kcal		
$F_G = fG$	8.4	$8.4 \cdot 10^{-18}$
$F_\mathfrak{S} = \sigma\mathfrak{S}$	$4.4 \cdot 10^{-7}$	$4.4 \cdot 10^{-19}$
$F_\mathfrak{S}/F_G$, %		5.2%
S, kcal/deg		
$S_G = sG$	0.162	$0.162 \cdot 10^{-18}$
$S_\mathfrak{S} = -\dfrac{d\sigma}{dT}\mathfrak{S}$	$0.147 \cdot 10^{-8}$	$0.147 \cdot 10^{-20}$
$S_\mathfrak{S}/S_G$, %		0.9%
U, kcal		
$U_G = uG$	52.4	$52.4 \cdot 10^{-18}$
$U_\mathfrak{S} = \left(\sigma - T\dfrac{d\sigma}{dT}\right)\mathfrak{S}$	$1.00 \cdot 10^{-6}$	$1.00 \cdot 10^{-18}$
$U_\mathfrak{S}/U_G$, %		1.9%
C, kcal/deg		
$C_G = c_\sigma G$	0.47	$0.47 \cdot 10^{-18}$
$C_\mathfrak{S} = -T\dfrac{d^2\sigma}{dT^2}\mathfrak{S}$	$0.11 \cdot 10^{-7}$	$0.11 \cdot 10^{-19}$
$C_\mathfrak{S}/G_G$, %		2.3%
Dimensions (area, volume) and weight of the system	$\mathfrak{S} = 314$ cm²$\quad V = 524$ cm³$\quad G = 524$ g	$\mathfrak{S} = 3.14 \cdot 10^{-10}$ cm²$\quad V = 0.524 \cdot 10^{-15}$ cm³$\quad G = 0.524 \cdot 10^{-9}$ g

al forces is negligible. How can such conditions arise? In the first
place, for small drops of liquid, the smaller the amount of material
the greater the surface-to-volume ratio and consequently the greater
the role of surface tension forces in comparison to the weight of
the material (which is proportional to the volume of the body). Actually it is well known from ordinary experience that drops of liquid
assume a spherical shape. In the second instance we have small
droplets in the atmosphere. As shown by photographs and movies,
when a droplet becomes so-to-speak weightless, it assumes a
spherical shape. Thirdly it is possible to have cases where the
effect of gravity is "switched off." The following experiment,
first performed in 1843, is well known. If we use olive oil and a
mixture of water and alcohol in proportions which make its density
equal to that of the oil, then according to Archimedes' law the weight
of the oil will equal the forces exerted by the mixture and the oil
becomes "weightless." Under these conditions the surface tension
forces cause the oil to assume the shape of a regular sphere (one
gets a sphere using oil in alcohol and not for alcohol in oil since
the surface tension of oil is greater than that of alcohol).

A number of specific effects which appear in thermodynamic
systems in a weightless state will be considered in detail in Chapter 8.

We should note yet another fact. Up to now we have spoken
only of the surface tension of a liquid. Surface tension also acts
at the boundary of a solid; however, its experimental measurement
is quite difficult since a solid phase does not flow.

6.5. Phase Equilibrium Conditions Taking Ac-
count of the Properties of the Phase Separation
Surface

We now consider the important question of the conditions
for phase equilibrium in an isolated system composed of two subsystems in different coexisting phases, taking account of the surface layer at the boundary between phases. (Earlier, in Section
2.4, we elucidated the conditions for phase equilibrium without considering the special properties of the phase separation surface.)

We analyze the phase equilibrium conditions in the present
isolated system by the same method as in Section 2.4. The basic

difference from the case treated in Section 2.4 is that we should now consider a system consisting not of two subsystems (in which the material is in different phases) but of three subsystems — subsystem 1, in which the material is in one phase; subsystem 2, containing material in the other phase; and the phase separation surface between subsystems 1 and 2, which we evidently should regard as an independent subsystem having specific properties differing from those of subsystems 1 and 2.

Since the system of interest is isolated, it is evident that

$$V_{syst} = \text{const},\tag{6.43}$$

$$G_{syst} = \text{const},\tag{6.44}$$

and

$$U_{syst} = \text{const}.\tag{6.45}$$

It should be stressed that a constant area of the phase separation surface ($\mathfrak{S} = \text{const}$) is not required of an isolated system; indeed, within the conditions (6.43) to (6.45) the area of the separation surface between phases can vary in an arbitrary way.

The surface layer between phases will be regarded as having no thickness (and consequently no volume); this simplification does not substantially affect the results of our analysis.

In view of what has been said, it is apparent that

$$V_{syst} = V_1 + V_2,\tag{6.46}$$

$$G_{syst} = G_1 + G_2,\tag{6.47}$$

$$U_{syst} = U_1 + U_2 + U_\sigma,\tag{6.48}$$

where the subscripts 1, 2, and σ refer, respectively, to subsystem 1, subsystem 2, and subsystem 3 (the surface layer).

From these relations and conditions (6.43) to (6.45) it follows that

$$dV_1 = -dV_2,\tag{6.49}$$

$$dG_1 = -dG_2,\tag{6.50}$$

$$dU_\sigma = -(dU_1 + dU_2). \tag{6.51}$$

We recall further than in an isolated system in equilibrium

$$dS_{\text{syst}} = 0. \tag{6.52}$$

The expression for the entropy of the system is written as

$$S_{\text{syst}} = S_1 + S_2 + S_\sigma. \tag{6.53}$$

Using (6.52), we obtain

$$dS_1 + dS_2 + dS_\sigma = 0. \tag{6.54}$$

The expressions for the differential entropy of each sub-system are written as follows. From Eq. (2.95) we find for subsystems 1 and 2 that

$$dS_1 = \frac{1}{T_1} dU_1 + \frac{p_1}{T_1} dV_1 - \frac{\varphi_1}{T_1} dG_1 \tag{6.55}$$

and

$$dS_2 = \frac{1}{T_2} dU_2 + \frac{p_2}{T_2} dV_2 - \frac{\varphi_2}{T_2} dG_2, \tag{6.56}$$

while for subsystem 3 we find from Eq. (6.4) that

$$dS_\sigma = \frac{1}{T_\sigma} dU_\sigma - \frac{\sigma}{T_\sigma} d\mathfrak{S} \tag{6.57}$$

(the term containing the chemical potential φ is absent in the latter equation because the term containing φ appears in the combined equation for the first and second laws of thermodynamics for a system with a variable amount of material; since we have assumed that the surface layer has no thickness, G = 0 for it, and thus dG = 0.

Substituting these values of dS_1, dS_2, and dS_σ into (6.54) and using (6.49) to (6.51), we find

$$\left(\frac{p_1}{T_1} - \frac{p_2}{T_2}\right) dV_1 - \left(\frac{\varphi_1}{T_1} - \frac{\varphi_2}{T_2}\right) dG_1 + \left(\frac{1}{T_1} - \frac{1}{T_\sigma}\right) dU_1 +$$

$$+ \left(\frac{1}{T_2} - \frac{1}{T_\sigma}\right) dU_2 - \frac{\sigma}{T_\sigma} d\mathfrak{S} = 0. \tag{6.58}$$

We analyze the relation we have found. The differentials dV_1, dG_1, dU_1, dU_2, and $d\mathfrak{S}$, which enter into this equation are independent of each other except for the differentials dV_1 and $d\mathfrak{S}$, which, as is easily seen, are related [a change in the volumes of the phases in general causes a change in the area of the phase separation surface so that in general $\mathfrak{S} = f(V_1)$]. We dwell in somewhat more detail on the independence of the differentials dU_1 and dU_2. Since we consider three subsystems, in principle we could think of a process where the energies of subsystems 1 and 3 or 2 and 3 change (remember that subsystem 3 is the surface dividing the phases) while the internal energies of, respectively, subsystems 2 and 1 remain unchanged.

It is evident that if the left-hand side of Eq. (6.58) is to be identically zero, the following equations must be satisfied:

$$\frac{1}{T_1} - \frac{1}{T_\sigma} = 0, \tag{6.59}$$

$$\frac{1}{T_2} - \frac{1}{T_\sigma} = 0, \tag{6.60}$$

$$\frac{\varphi_1}{T_1} - \frac{\varphi_2}{T_2} = 0, \tag{6.61}$$

$$\left(\frac{p_1}{T_1} - \frac{p_2}{T_2} \right) dV_1 - \frac{\sigma}{T_\sigma} d\mathfrak{S} = 0. \tag{6.62}$$

It follows from (6.59) and (6.60) that

$$T_1 = T_2 = T_\sigma, \tag{6.63}$$

while it follows from (6.61), using (6.63), that

$$\varphi_1 = \varphi_2. \tag{6.64}$$

Thus we have established that if the two phases are in equilibrium and we consider the properties of the surface layer then the temperatures and chemical potentials of the coexisting phases are mutually equal. This result agrees with Eqs. (2.120) and (2.141) obtained in Chapter 2; consequently when applied to these conditions consideration of the surface layer does not introduce any new results.

It is quite different with the third equilibrium condition for coexisting phases. We recall that previously when the surface layer was not taken into account we concluded that the pressure is the same [Eq. (2.121)] in the coexisting phases. It is otherwise in the present case. Using (6.63), Eq. (6.62) assumes the form

$$p_1 = p_2 + \sigma \frac{d\mathfrak{S}}{dV_1}. \tag{6.65}$$

from which it is evident that the pressure in the coexisting phases differs by the amount $\sigma(d\mathfrak{S}/dV_1)$, where $d\mathfrak{S}/dV_1$ is the derivative of the surface area with respect to the volume. Here, since σ is always positive, the question of which of the two coexisting phases has the higher pressure is determined by the sign of the derivative $d\mathfrak{S}/dV_1$.[†]

As we know from analytic geometry,

$$\frac{d\mathfrak{S}}{dV} = \frac{1}{\rho_1} + \frac{1}{\rho_2}, \tag{6.66}$$

where ρ_1 and ρ_2 are the principal radii of curvature of the surface. Substituting (6.66) into (6.65), we find that

$$p_1 = p_2 + \sigma \left(\frac{1}{\rho_1} + \frac{1}{\rho_2} \right). \tag{6.67}$$

In the particular case of a spherical separation surface ($\rho_1 = \rho_2$), Eq. (6.67) is transformed in the following way:

$$p_1 = p_2 + \frac{2\sigma}{\rho}; \tag{6.68}$$

Equation (6.67) is called the Laplace equation. Here we have established the Laplace equation by thermodynamic methods. This equation can be obtained without difficulty in a purely mechanical way by treating the surface layer as an elastic film.

For this purpose we consider a portion of the phase separation as shown in Fig. 6.6. The arrows in the figure show the surface tension forces. As previously noted, forces act along the perimeter of the region, along the normal to the perimeter, and along the tangent to the surface. For a planar phase separation surface at all points on the

---

[†]In view of (6.49) it is clear that

$$\frac{d\mathfrak{S}}{dV_1} = -\frac{d\mathfrak{S}}{dV_2}.$$

perimeter of the region (Fig. 6.6a) the surface tension forces lie in the same plane so that apparently the sum of these forces is zero. It is different for a curved interphase surface. As we see from Fig. 6.6b, in this case the forces lie in different planes at different points of the perimeter of the region and the total force is not zero here. It is apparent that this total force is greater the greater the curvature of the surface separating the phases. It is not hard to see that this total force is directed toward the concave side of the surface. Thus the phase which is on the concave side of the surface is subject to an additional pressure due to surface tension.

We must explain how this force is related to the magnitude of the surface tension and the curvature of the surface. We examine a small square of a curvilinear surface (Fig. 6.7) and calculate the total surface tension force acting along the perimeter of this square. We first consider the forces acting on parts AB and CD of the perimeter. The force acting along AB is

$$\mathfrak{F}_{AB} = \sigma l_{AB},\tag{6.69}$$

where l_{AB} is the length of segment AB, and the corresponding force acting along CD is

$$\mathfrak{F}_{CD} = \sigma l_{CD}.\tag{6.70}$$

The total of the forces $\mathfrak{F}_{AB}$ and $\mathfrak{F}_{CD}$ is found as the sum of the projections of the vectors $\mathfrak{F}_{AB}$ and $\mathfrak{F}_{CD}$ on the line OO' normal to the surface region of interest at its center. This total force is

$$\mathfrak{F}_{AB} \sin \alpha_{I} + \mathfrak{F}_{CD} \sin \alpha_{II},$$

where α_1 is the angle between the normal OO' and the radius of curvature ρ_I extended to point I of the surface region of interest in a plane perpendicular to the sides AB and CD.

Since

$$l_{AB} = l_{CD} = l,\tag{6.71}$$

it is evident that in absolute value

$$|\mathfrak{F}_{AB}| = |\mathfrak{F}_{CD}|\tag{6.72}$$

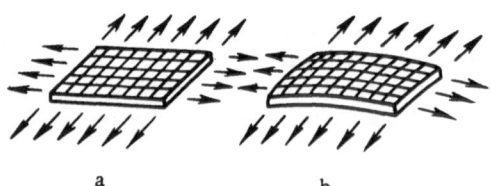

a   b

Fig. 6.6

and consequently the sum of the forces $\mathfrak{F}_{AB}$ and $\mathfrak{F}_{CD}$ will be

$$2\sigma l \sin \alpha_{\mathrm{I}}.$$

In a similar way the total of the forces $\mathfrak{F}_{AD}$ and $\mathfrak{F}_{BC}$ will be $2\sigma l \sin \alpha_{\mathrm{II}}$, where α_{II} is the angle between the normal OO' and the radius of curvature ρ_{II} extended to point II of the surface region of interest in the plane perpendicular to the sides AD and BC.

Consequently the total of the forces $\mathfrak{F}_{AB}$, $\mathfrak{F}_{BC}$, $\mathfrak{F}_{CD}$, and $\mathfrak{F}_{AD}$, i.e., the whole surface tension force acting on the surface region of interest, is

$$\mathfrak{F} = 2\sigma l \,(\sin \alpha_{\mathrm{I}} + \sin \alpha_{\mathrm{II}}). \tag{6.73}$$

As we see from Fig. 6.7,

$$\sin \alpha_{\mathrm{I}} = \frac{l/2}{\rho_{\mathrm{I}}} \quad \text{and} \quad \sin \alpha_{\mathrm{II}} = \frac{l/2}{\rho_{\mathrm{II}}}. \tag{6.74}$$

In view of these relations, Eq. (6.73) assumes the form

$$\mathfrak{F} = \sigma l^2 \left(\frac{1}{\rho_{\mathrm{I}}} + \frac{1}{\rho_{\mathrm{II}}} \right). \tag{6.75}$$

Dividing $\mathfrak{F}$ by the surface area of the region of interest (this area, as can easily be seen, is l^2), we find the magnitude of the additional pressure arising from surface tension forces:

$$p^* = \sigma \left(\frac{1}{\rho_{\mathrm{I}}} + \frac{1}{\rho_{\mathrm{II}}} \right). \tag{6.76}$$

Further, the additional pressure due to surface tension forces is equal to the pressure difference between the coexisting phases:

$$p^* = p_1 - p_2, \tag{6.77}$$

i.e., Eq. (6.76) agrees with Eq. (6.67).

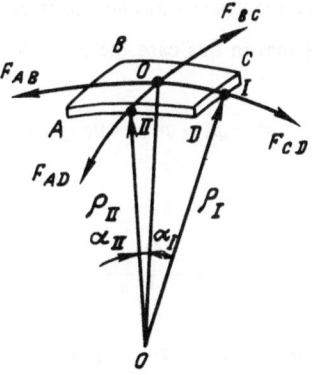

Fig. 6.7

It follows from Eqs. (6.67) and (6.68) that for a planar phase separation surface ($\rho = \infty$)

$$p_1 = p_2. \tag{6.78}$$

Thus for a planar phase surface, consideration of the surface layer does not change the condition (2.121) for phase equilibrium — the pressures in the coexisting phases are the same. For a curvilinear surface separating the phases there is a pressure difference given by Eq. (6.67).

It is undoubtedly of interest how the pressure changes in coexisting phases if the surface separating them, which was previously planar, becomes curved for any reason.

The pressure in the coexisting phases with a planar surface separating them will be called p.

As we know from general thermodynamics, when the pressure increases in one of the coexisting phases the pressure also rises in the second phase in agreement with the so-called Poynting equation

$$\left(\frac{\partial p_2}{\partial p_1}\right)_T = \frac{v_1}{v_2}, \tag{6.79}$$

where the subscripts 1 and 2 refer to the first and second coexisting phases. In other words, if the pressure in the first phase increases from p to p_1, i.e., by an amount

$$\Delta p_1 = p_1 - p, \tag{6.80}$$

then the pressure in the second phase also increases from p to p_2, i.e., by an amount

$$\Delta p_2 = p_2 - p; \tag{6.81}$$

and the relation between the quantities Δp_1 and Δp_2 is given by Eq. (6.79).

It is evident from this that in this case the pressure difference between the coexisting phases is

$$p^* = p_1 - p_2 = \Delta p_1 - \Delta p_2. \tag{6.82}$$

In agreement with the Poynting law

$$\Delta p_2 = \int_p^{p_1} \frac{v_1}{v_2} \, dp_1. \tag{6.83}$$

If the ratio v_1/v_2 does not vary strongly with a change of p_1 (this condition is satisfied in most cases), then (6.83) can, to a good approximation, be written in the following form:

$$\Delta p_2 = \frac{v_1}{v_2} \Delta p_1. \tag{6.84}$$

We find from Eq. (6.82), using (6.84),

$$\Delta p_1 = \frac{v_2}{v_2 - v_1} p^* \tag{6.85}$$

and

$$\Delta p_2 = \frac{v_1}{v_2 - v_1} p^*, \tag{6.86}$$

whence

$$p_1 = p + \frac{v_2}{v_2 - v_1} p^* \tag{6.87}$$

and

$$p_2 = p + \frac{v_1}{v_2 - v_1} p^*. \tag{6.88}$$

If the excess pressure in one of the phases arises from curvature of the separation surface then it is obvious that the pressure difference $p^* = p_1 - p_2$ in the coexisting phases will be given by the Laplace equation (6.67), i.e.,

$$p^* = \sigma \left(\frac{1}{\rho_1} + \frac{1}{\rho_2} \right).$$

In this case we have

$$p_1 = p + \frac{v_2 \sigma}{v_2 - v_1} \left(\frac{1}{\rho_1} + \frac{1}{\rho_2} \right) \tag{6.89}$$

and

$$p_2 = p + \frac{v_1 \sigma}{v_2 - v_1} \left(\frac{1}{\rho_1} + \frac{1}{\rho_2} \right). \tag{6.90}$$

6.6. Capillarity

It is well known from common experience that if we place one end of a glass tube of small diameter (a capillary) in water the water rises into the capillary to a height which is higher than the common level of the liquid. This phenomenon (it is usually called capillarity) can be explained by the action of surface tension forces. We will treat it in more detail.

The surface of a liquid in a vessel always curves near the walls. The nature of the surface curvature of the liquid at the walls of the vessel can be explained as follows. As noted earlier, surface

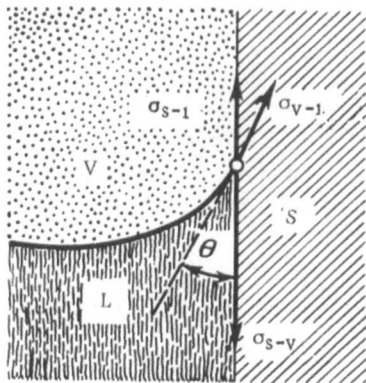

Fig. 6.8

tension affects not only the liquid—vapor interface but also separation surfaces of solid and liquid phases and solid and gaseous phases.
It is clear from this that three forces act at the point where the
surface of a liquid meets the walls of the vessel. There is not only
a surface tension force between the liquid and its vapor (which we
denote by σ_{1-v}) but also a surface tension force between the liquid
and the solid surfaces of the vessel walls (σ_{1-s}) and a surface tension force between the vapor and the solid surface of the vessel
walls (σ_{v-s}). It is clear that the shape of the liquid surface will
be in equilibrium when the total of these three forces is zero (Fig.
6.8). The force σ_{v-s} points down along the wall of the vessel, the
force σ_{1-s} acts up along the wall of the vessel, and thus these forces
are not only fixed in value (their values are determined uniquely
by the surface tension coefficients for the given combinations of
phases) but also in direction. As for the force σ_{1-v}, it is fixed in
value but its direction will be determined by the angle at which the
liquid surface meets the wall of the vessel. From what was said
above it is clear that this angle should be such that the sum of the

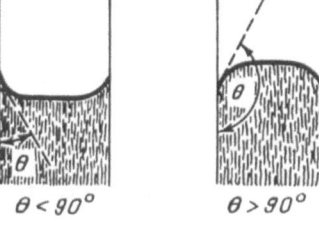

$\theta < 90°$ $\theta > 90°$

Fig. 6.9

TABLE 6.3

Liquid—solid	θ	Liquid—solid	θ
Water—glass[*]	0°	Acetic acid—glass	20°
Methyl alcohol—glass	0°	Benzene—glass	0°
Ethyl alcohol—glass	0°	Parafin	
ether—glass	16°	oil—glass	26°
		Turpentine—glass	17°

[*]With completely clean glass.

three forces of interest is zero, i.e., that, as we see from Fig. 6.8, the following condition is satisfied

$$\sigma_{1-s} + \sigma_{1-v} \cos \theta - \sigma_{v-s} = 0. \qquad (6.91)$$

The angle between the liquid surface and the solid surface is called the contact angle (θ in Fig. 6.8). The contact angle is determined by the properties of the liquid and its saturated vapor and the properties of the solid material in the vessel wall. It is meaningless to speak of the contact angle simply for a given liquid; one must indicate the solid material the liquid is in contact with. The contact angle θ is thus a constant, characteristic of the interaction of the liquid with the solid.

The angle θ can be different for different liquids in combination with different solids and can vary from 0 to 180° (Fig. 6.9). If $\theta < 90°$ the liquid surface is concave; in this case we say that the liquid wets the surface of the given solid. If $\theta > 90°$ the liquid surface is convex (the liquid does not wet the surface). As well-known examples we can cite, for example, for a glass surface, water, which wets glass, and mercury, which does not wet glass. The contact angles for various materials are given in Table 6.3.

We also consider what happens with a liquid in a capillary. At the perimeter of the liquid surface there are surface tension forces which point, as always, along the tangent to this surface. Since the surface of the liquid is curved, the sum of these forces[†]

[†]We do not give the resultant of the three forces (σ_{v-1} σ_{v-s}, and σ_{1-s}) discussed above, which determines the shape of the liquid surface and is applied at the point where the liquid touches the wall of the vessel.

is nonzero. This sum, acting at the center of the liquid surface,
is to be understood to be directed along the normal to the concave side of the surface. The total surface tension force acting on
the perimeter of the liquid in a round tube is

$$F_\sigma = \sigma \cdot 2\pi b \cos \theta, \tag{6.92}$$

where b is the radius of the capillary.

Thus, at the surface of a liquid in a vessel there is a vertical
upward force F_σ for liquid which wet the solid surface ($\cos \theta > 0$).
The effect of this force also causes the liquid to rise in the capillary.
To what height can the liquid rise in the capillary? It is clear that
this rise will continue as long as the force F_σ exceeds the weight of
the column of liquid in the capillary which has risen above the common level (in an atmosphere of the saturated vapor of the liquid).
If we denote the height of the column by h, then its weight will be†

$$G = \frac{\pi b^2 h}{v_1 - v_\mathbf{v}}. \tag{6.93}$$

Equating the right-hand sides of Eqs. (6.92) and (6.93), we
find that the height to which the liquid rises in the capillary is

$$h = \frac{\sigma \cos \theta \, (v_1 - v_\mathbf{v})}{b}. \tag{6.94}$$

It follows from Eq. (6.94) that the height the liquid rises
in the capillary is greater the less the capillary radius, the greater
the surface tension, and the smaller the contact angle. Furthermore it follows from Eq. (6.94) that h is negative for liquids which
do not wet the surface ($\theta > 90°$ and $\cos \theta < 0$). This conclusion
is not necessarily paradoxical; indeed if, for example, we place a
glass capillary in mercury the level of mercury in the capillary
turns out to be lower than the level in the vessel.

---

†For simplicity we neglect the curved surface of the liquid in the capillary and treat
the column of liquid in the capillary as a cylinder.

CHAPTER 7

Gases and Liquids in a Gravitational Field

7.1. Basic Thermodynamic Relations for a System in a Gravitational Field

In solving a number of problems of important practical value a substantial role is played by the effect of gravitation forces† on the thermodynamic properties of the system.

As is well known, in a column of a gas or liquid the pressure varies with height due to the change of the hydrostatic pressure in the column of gas (liquid) with height. We consider an element of a gas column having height dh (Fig. 7.1). It is evident that the pressure in the plane b–b will be greater than the pressure in the plane a–a since in a gas the plane b–b receives the additional weight of the gas which lies between the planes a–a and b–b. It is further obvious that the pressure change in an element of the gas column of height dh is

$$dp = \frac{\gamma dV}{\mathfrak{S}}, \tag{7.1}$$

where γ is the specific weight of the gas in the element of the column, dV is the elementary volume, and $\mathfrak{S}$ is the cross-sectional area of an element of the gas column. Since $dV = \mathfrak{S}\,dh$, we find from (7.1)

---

†Everywhere in this chapter we consider the gravitational field of the earth (free acceleration g = 9.8 m/sec²). If a gravitational field with a different free acceleration is assumed, then all of the calculations must be done with a value of g corresponding to that field.

171

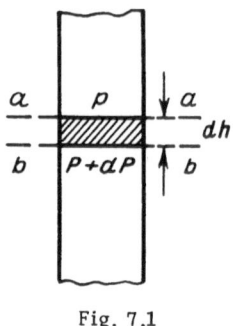

Fig. 7.1

that

$$dp = -\gamma dh. \tag{7.2}$$

In view of the fact that $\gamma = 1/v$, where v is the specific volume, we find that

$$dp = -\frac{dh}{v} \tag{7.3}$$

or, what is the same,

$$\left(\frac{\partial p}{\partial h}\right)_T = -\frac{1}{v}. \tag{7.3a}$$

The minus sign shows that with increasing height (dh > 0) the pressure of the gas (liquid) decreases (dp < 0).

The effect of gravitational forces on the state of the thermodynamic system (which can be, for example, the gas or liquid in a vessel) manifests itself primarily in a change of pressure with height.[†] With vessels of moderate height this change as a rule is negligible in relation to the absolute pressure in the vessel, and consequently the effect of a change of p with height is negligible in most cases. Therefore the effect of gravitation is frequently neglected. However for those states of a material in which the compressibility of the material is large, even an insignificant change in the pressure with height in the vessel will lead to a significant change in the density and other thermodynamic properties of the

---

[†] Therefore the gravitational field only fails to significantly affect the thermodynamic properties of a system when its height is small. The necessary smallness of height is different for different processes and states of the material.

material. Speaking of states in which the compressibility of the material is quite large, we first of all have in mind the near-critical region; we recall that at the critical point itself the isothermal compressibility of a pure material is infinite:

$$\left(\frac{\partial v}{\partial p}\right)_T^{cr} = \infty.$$ (7.4)

This fact must always be taken into account in carrying out experimental investigations of the thermodynamic properties of materials at the critical point and near it. Neglect of this effect can lead to erroneous conclusions. Therefore at present no investigation of the properties of a material near the critical point which pretends to a high degree of precision can be done without considering the effect of gravitation on the experimental results.

In a number of cases it is important to consider the change of pressure with height of the column also for other states which have extremal values of the derivatives of other thermodynamic quantities (such for example, as the degree of dissociation and the degree of ionization of a gas) with respect to pressure at T = const.

It is also necessary to consider gravity when the compressibility of the system may not be too large but the height of the gas (liquid) column of interest is substantial (for example, in analyses of the state of the atmosphere).

The elementary work needed to move a weight G a height dh in a gravitational field is

$$dL^* = G\,dh.$$ (7.5)

Consequently, in agreement with (1.27a), the combined equation of the first and second laws of thermodynamics for a system in a gravitational field is written as follows:

$$TdS = dU + p\,dV + G\,dh.$$ (7.6)

It is interesting to note, as seen from (7.5), that for work done by a system in a gravitational field the generalized force – the weight G of the body – in contrast to other generalized forces known to us, is not an intensive quantity but an extensive one, while the generalized coordinate, the height h, on the contrary is not extensive but intensive. We note in this connection that Eq. (7.6) can be transformed to a more "usual" (with regard to intensiveness and the capacity factor) form; we use the Legendre transformation

$$Gdh = d\,(Gh) - hdG, \qquad (7.7)$$

and find from (7.6) that

$$TdS = d\,(U + Gh) + pdV - hdG. \qquad (7.8)$$

The sum (U + Gh) which enters this equation can be interpreted as the "total" energy of the system of interest,

$$U^* = U + Gh, \qquad (7.9)$$

by analogy with the "total" enthalpy of such a system defined in general by Eq. (2.39) and applied to the given type of system considered below.

If the weight G of the system remains unchanged then Eq. (7.6) written for weight-specific values assumes the form

$$Tds = du + pdv + dh. \qquad (7.10)$$

Furthermore, in agreement with (2.38) and (2.39), it is clear that for a system in a gravitational field the enthalpy I^* of the whole system is

$$I^* = U + pV + Gh \qquad (7.11)$$

or

$$I^* = I + Gh, \qquad (7.12)$$

where Gh is the potential energy of a body of weight G raised to a height h. The corresponding weight-specific enthalpy for such a system is

$$i^* = i + h. \qquad (7.13)$$

The second term in (7.13) characterizes the potential energy accumulated by a body raised to a height h. Thus if a body is raised to a height of 1000 m it is able to do an amount of work equal to 1000 kg · m/kg = 2.34 kcal/kg. This "correction" to the enthalpy is determined only by height and does not depend on any other state parameters of the system. Therefore, for example, the heat capacity c_p^* of material in a gravitational field is given by the obvious relation†

$$c^*_p = \left(\frac{\partial i^*}{\partial T}\right)_p,$$ (7.14)

which equals the usual heat capacity c_p since h does not depend on T. Thus the heat capacity c_p does not depend on the presence of a gravitational field although the enthalpy does depend on h‡

We should make one important remark. The specific quantities of interest (the specific volume, specific enthalpy, specific heat capacity, etc.) are everywhere referred to a unit weight of the body (1 kg or 1 g). But in other gravitational fields (for example, on the moon, where g is one-sixth that on the earth) referral of specific values to unit weight encounters a number of difficulties (for example, the total heat capacity of a piece of metal will undoubtedly be the same on the earth as on the moon but the specific heat capacities of this metal per unit weight on the moon will be six times greater than on the earth) and under weightless conditions referral to unit weight is generally useless. Evidently from this viewpoint it is more correct in principle to refer specific values not to unit weight but to unit mass of a body since the amount of mass in a body is the same in any gravitational field (it is unchanged as long as the rate of motion of the body is substantially less than the velocity of light). For ordinary earth conditions both of these methods for defining specific values are the same, since in the earth's gravitational field (g = 9.8 m/sec^2) the weight of a body is numerically equal to its mass.

If a system is in a gravitational field and the compressibility of the material is quite large, then under static conditions it is practically impossible to ensure constant pressure over the height of the system. Therefore in treating systems in a gravitational field we will not consider isobaric conditions – this case makes no physical sense. In what follows we consider a column of gas (liquid) in a gravitational field and we will consider only isothermal and adiabatic conditions.

---

†Equation (7.14) is identical to one found previously in Chapter 1, Eq. (1.36)

$$c^*_p = c_p + \left(\frac{\partial l^*}{\partial T}\right)_p.$$

‡In fact this is only valid for systems whose dimensions are small. Otherwise the idea of a heat capacity c_p under constant pressure conditions for the system makes no sense as it cannot be realized in the system.

7.2. Distribution of Pressure and Other Quantities with Height of a Gas or Liquid Column

For a number of problems of practical importance it is of interest to seek the distributions of pressure, specific volume, and other quantities along the height of a column of gas or liquid.

We find the pressure distribution over the height of an isothermal column (vessel).

It is clear from Eq. (7.3) that the pressure distribution with respect to vessel height is determined by the relation

$$p(h) = p_0 - \int_0^h \frac{dh}{v},\tag{7.15}$$

where p(h) is the pressure in a cross section of the vessel at height h from the bottom of the vessel, and p_0 is the pressure at the bottom of the vessel.

In order to use Eq. (7.15) to find the pressure distribution over the height of the vessel we must know the specific volume in each cross section over the whole height of the vessel. However, since the specific volume along an isotherm depends uniquely on the pressure, the distribution of v with respect to height is itself a function of the required pressure distribution. It is thus evident that to solve this problem we need an empirical equation of state for the given material in the form v = f(p), but finding the relation p(h) then also involves a number of serious difficulties in the majority of cases. Alternatively, we can use the method of successive approximations (for this we must have experimental or numerical values calculated from the equation of state or the p—v relation for the material of interest along the isotherm used).

This calculation is done as follows. In the first approximation we assume that the specific volume is constant over the height of the vessel and equal to the specific volume with a pressure equal to the pressure at the lowest point of the vessel (p_0), i.e., v(h) = v_0 = const; it then follows from (7.15) that

$$p_1(h) = p_0 - \frac{h}{v_0}.\tag{7.16}$$

The distribution of pressure over the height of the vessel found in this way can be used to find the distribution of specific volume over the height of the vessel, $v_1(h)$ with the aid of data on the p–v relation for the isotherm of interest. Having the relation $v_1(h)$, we can use Eq. (7.15) to find the distribution of pressure with respect to the height of the vessel in the second approximation:

$$p_2(h) = p_0 - \int_0^h \frac{dh}{v_1(h)}. \qquad (7.17)$$

Knowing the distribution $p_2(h)$, we can use data on the p–v relation to find the distribution of specific volume over the height of the vessel in the second approximation, $v_2(h)$, and then repeat this procedure until good convergence is obtained (i.e., until the results of successive approximations are nearly equal). In practice, the second approximation usually gives good convergence.[†]

Figure 7.2 shows the distribution of the specific volume of water over the height of a vessel 1 m high whose temperature equals the critical temperature of water (374.15°C). As we see from this figure, the specific volume changes quite significantly with the height of the vessel (by approximately 15%). The inflection point in the v(h) curve in Fig. 7.2 corresponds to the height in the vessel where the pressure equals the critical pressure (for water, p_{cr} = 225.65 kg/cm^2); consequently the specific volume at this point is v_{cr}.

Knowing the distribution of pressure and specific volume over height and the dependence of any other thermodynamic or caloric quantity on p or v along the isotherm of interest, it is not hard to find the distribution of this quantity over the height of the vessel. For example, Fig. 7.3 shows the distribution of the heat capacity c_V over the height of a vessel 1 m high filled with water at 374.15°C constructed using the experimentally measured relation $c_V = f(v)$ along the critical isotherm and the distribution of specific volume with respect to height given in Fig. 7.2.

If the pressure and temperature of a gas are such that it can be regarded as ideal (for example, atmospheric air), then for the ideal gas

$$v = \frac{RT}{p}, \qquad (7.18)$$

___

[†]Since the dependence of pressure on height is most often nearly linear for systems of relatively low height.

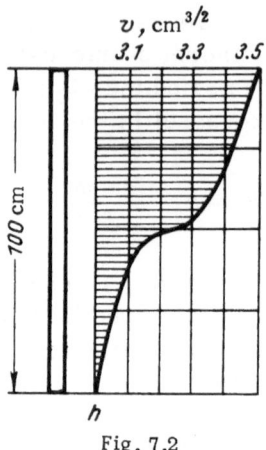

Fig. 7.2

and Eq. (7.3) assumes the following form:

$$dp = -\frac{p}{RT}\, dh,$$

$$(7.19)$$

whence

$$\frac{dp}{p} = -\frac{dh}{RT}.$$

$$(7.20)$$

Integrating, we find

$$\ln\frac{p(h)}{p_0} = -\frac{1}{R}\int_{h_0}^{h}\frac{dh}{T}.$$

$$(7.21)$$

If we consider an isothermal column of gas then T is constant and can be removed from under the integral sign:

$$\ln\frac{p(h)}{p_0} = -\frac{h - h_0}{RT}.$$

$$(7.22)$$

Thus we find the following formula for the distribution of pressure in an isothermal column of an ideal gas (the so-called barometric formula):

$$p(h) = p_0 e^{-\frac{h - h_0}{RT}}.$$

$$(7.23)$$

As seen from this relation, the relation p(h) is of an exponential nature for the case of interest. The pressure distribution (the

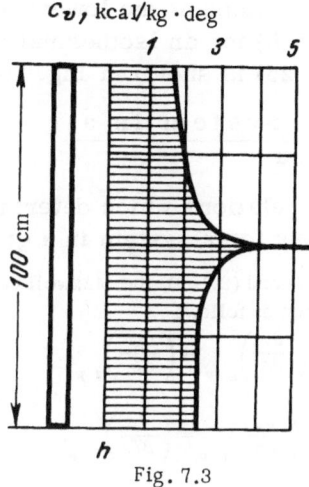

Fig. 7.3

quantity p/p_0) over the height of an isothermal air column 1000 m
high is shown in Fig. 7.4.

As for the distribution of specific volume over the height
of an isothermal column of an ideal gas, from the equation of state
of the ideal gas (7.18), using (7.23), we find

$$v(h) = \frac{RT}{p_0} e^{\frac{h-h_0}{RT}} \qquad (7.24)$$

or, what is the same,

$$v(h) = v_0 e^{\frac{h-h_0}{RT}}, \qquad (7.25)$$

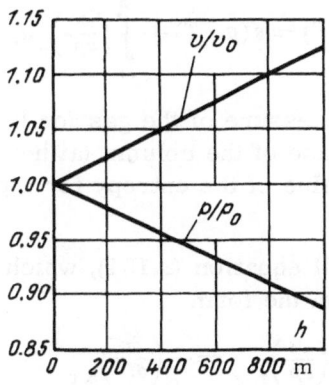

Fig. 7.4

where v_0 is the specific volume at the lowest point of the column ($h = h_0$). The nature of $v(h)$ for an isothermal column of an ideal gas described by Eq. (7.25) is shown in Fig. 7.4.

7.3. Entropy of a System in a Gravitational Field

We formulate the relations which determine the height, h dependence of the entropy for a system in a gravitational field.

In agreement with (2.147) and (2.150), the Maxwell equation for a system in a gravitational field is written as follows

$$\left(\frac{\partial s}{\partial h}\right)_{g,\,p} = \left(\frac{\partial g}{\partial T}\right)_{s,\,p}, \tag{7.26}$$

$$\left(\frac{\partial s}{\partial h}\right)_{T,\,v} = \left(\frac{\partial g}{\partial T}\right)_{h,\,v}. \tag{7.27}$$

Since in fact the gravitational constant g does not depend on the temperature of the medium, it is clear from these equations that the entropy of a system in a gravitational field does not depend on its height. For what systems is this conclusion valid? Evidently only for those system where Eqs. (7.26) and (7.27) are valid. But are not these equations perhaps valid for all systems in a gravitational field? No, not for all of them. The fact is that both partial derivatives entering Eq. (7.26) were taken with p = const while both derivatives in Eq. (7.27) were taken with v = const. These conditions are satisfied only for systems of small height (p and v nearly unchanged over the height of the system). For a column of liquid or gas there are changes in the pressure p and specific volume v with h, so that Eqs. (7.26) and (7.27) are inapplicable for such systems.

We can calculate the specific entropy of a liquid or gas in an isothermal column of height h using the obvious relation

$$s(p,\,T) = s(p_0,\,T) + \int_{p_0}^{p} \left(\frac{\partial s}{\partial p}\right)_T dp. \tag{7.28}$$

Here p and p_0 are the pressure of the gas (or liquid), respectively, at height h and at the base of the column (at height h_0), while $s(p_0, T)$ is the known value of the entropy for a pressure p_0 (i.e., with $h = h_0$).

Using the Maxwell equation (2.154), which when applied to our system is written in the form

$$\left(\frac{\partial s}{\partial p}\right)_{T,\,g} = -\left(\frac{\partial v}{\partial T}\right)_{p,\,g}. \tag{7.29}$$

or, what is the same,

$$\left(\frac{\partial s}{\partial p}\right)_T = -\left(\frac{\partial v}{\partial T}\right)_p, \tag{7.29a}$$

Eq. (7.28) transforms as follows:

$$s(p,\,T) = s(p_0,\,T) - \int_{p_0}^{p} \left(\frac{\partial v}{\partial T}\right)_p dp. \tag{7.30}$$

As shown above, we can find the distribution of pressure over the height of the isothermal column of interest. Furthermore, using the equation of state or numerical methods, we can find the value of $(\partial v/\partial T)_p$ corresponding to given p and T for any value of p from p_0 to p (corresponding to a height interval from h_0 to h). Knowing these values we can easily calculate the integrals which enter the right-hand side of Eq. (7.30); this calculation can be done graphically or numerically. As we see from (7.30), the entropy of a gas (liquid) in an isothermal column usually increases with height; indeed since $(\partial v/\partial T)_p$ as a rule[†] is positive and p < p_0, we find

$$s(p,\,T) > s(p_0,\,T). \tag{7.31}$$

When the isothermal column of interest is an ideal gas the value of this integral can be determined analytically. From Eq. (7.18) it follows that for an ideal gas

$$\left(\frac{\partial v}{\partial T}\right)_p = \frac{v}{T} = \frac{R}{p}. \tag{7.32}$$

In view of this relation we find from (7.30)

$$s(p,\,T) = s(p_0,\,T) - R \ln \frac{p}{p_0}. \tag{7.33}$$

Substituting values of p(h) into this equation from Eq. (7.23), we

---

[†]In some cases the magnitude of $(\partial v/\partial T)_p$ can change sign. Thus, for example, for water$(\partial v/\partial T)_p$ turns out to be negative at temperatures below 3.98°C (the temperature of greatest density). Consequently in an isothermal column of water with temperatures $0 \le T \le 3.98°C$ the entropy decreases with height while at 3.98°C the entropy hardly varies with height.

find

$$s(p, T) = s(p_0, T) + \frac{h - h_0}{T} \qquad (7.34)$$

or, what is the same,

$$s(h, T) = s(h_0, T) + \frac{h - h_0}{T}. \qquad (7.34a)$$

As we see from this relation, the entropy increases linearly with height in an isothermal column of an ideal gas.

Figure 7.5 shows the nature of the relation between the entropy of an ideal gas and the height for different isotherms. As we see from (7.34), the angular coefficient of such a linear isotherm is $1/T$. This conclusion can also be reached as follows. From the obvious relation

$$\left(\frac{\partial s}{\partial h}\right)_T = \left(\frac{\partial s}{\partial p}\right)_T \left(\frac{\partial p}{\partial h}\right)_T, \qquad (7.35)$$

by substituting $(\partial s/\partial p)_T$ in agreement with Eqs. (7.29a) and (7.32),

$$\left(\frac{\partial s}{\partial p}\right)_T = -\left(\frac{\partial v}{\partial T}\right)_p = -\frac{v}{T}, \qquad (7.36)$$

while we find for $(\partial p/\partial h)_T$, in agreement with Eq. (7.3a),

$$\left(\frac{\partial p}{\partial h}\right)_T = -\frac{1}{v},$$

that

$$\left(\frac{\partial s}{\partial h}\right)_T = \frac{1}{T}. \qquad (7.37)$$

from which (7.34) follows automatically.

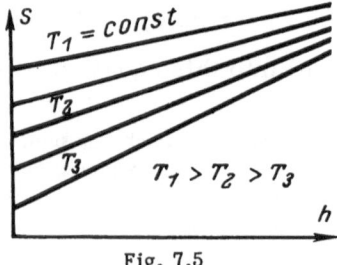

Fig. 7.5

As we see from (7.37), the higher the temperature the higher lie the isotherms in the s—h diagram.

In particular we see from this that for certain values of h two isotherms of an ideal gas can intersect in the s-h diagram. This means that at their intersection point the entropy difference of these two isotherms

$$s(h, T_2) - s(h, T_1) = \int_{T_1}^{T_2} \left(\frac{\partial s}{\partial T}\right)_h dT \tag{7.38}$$

will be zero for this h. Evidently this is only possible when $(\partial s/\partial T)_h$ changes sign. In what cases is this condition satisfied?

In agreement with (1.49) the derivative $(\partial s/\partial T)_h$ can be put in the form

$$\left(\frac{\partial s}{\partial T}\right)_h = \left(\frac{\partial s}{\partial T}\right)_p + \left(\frac{\partial s}{\partial p}\right)_T \left(\frac{\partial p}{\partial T}\right)_h, \tag{7.39}$$

whence, using (1.34), we have

$$\left(\frac{\partial s}{\partial T}\right)_h = c_p + T \left(\frac{\partial s}{\partial p}\right)_T \left(\frac{\partial p}{\partial T}\right)_h. \tag{7.40}$$

In agreement with the Maxwell equation (2.154) we have

$$\left(\frac{\partial s}{\partial p}\right)_T = -\left(\frac{\partial v}{\partial T}\right)_p,$$

and using the Clapeyron equation we have

$$\left(\frac{\partial s}{\partial p}\right)_T = -\frac{R}{p}. \tag{7.41}$$

We find from (7.23) for an isothermal column of an ideal gas (assuming that p_0 does not vary with temperature)

$$\left(\frac{\partial p}{\partial T}\right)_h = \frac{p_0 h}{RT^2} e^{-\frac{h}{RT}} = \frac{ph}{RT^2}, \tag{7.42}$$

and substituting (7.41) and (7.42) into (7.40), we find

$$\left(\frac{\partial s}{\partial T}\right)_h = c_p - \frac{h}{T},$$

whence it is clear that $(\partial s/\partial T)_h$ changes with sign when $h = c_p/T$.

7.4. Adiabatic Flow in a Gravitational Field

Above we considered an isothermal column of gas. An adiabatic column of gas (or liquid) is not of interest in itself since in agreement with the discussion of Chapter 2, in an adiabatic isolated system with constant volume the temperature is the same in all parts of the system; thus this case reduces to that of an isothermal column, discussed above.

However, threre is great practical interest in analyzing adiabatic flow of a gas or liquid in a gravitational field.

We recall that, as shown in Chapter 1, the differential equation for flow in a gravitational field takes the following form:

$$dq_{ext} = di + \frac{wdw}{g} + dh + dl_{tech} \tag{1.12}$$

This equation is valud for flows both with and without friction.

For adiabatic flow ($q_{ext} = 0$) there is no technical work ($l_{tech} = 0$), and this equation assumes the form

$$di + \frac{wdw}{g} + dh = 0. \tag{7.43}$$

If the velocity is small enough that wdw/g can be neglected, then for adiabatic flow of a gas or liquid in a gravitational field we find from (7.43) that

$$di + dh = 0, \tag{7.44}$$

i.e.,

$$di = -dh. \tag{7.44a}$$

Integrating this equation from a flow point at height h_1 to a point at height h_2 we find

$$i_1 - i_2 = h_2 - h_1. \tag{7.45}$$

This relation shows that the enthalpy decreases by Δh for the case of interest (adiabatic flow with a comparatively low velocity) in raising a liquid (gas) by a height Δh = $h_2 - h_1$.

Equation (7.44a) is the basis of the original method for the experimental determination of the heat capacity c_p of a gas at low pressures. A calorimeter for use with this method is shown

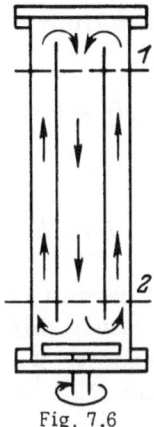

Fig. 7.6

schematically in Fig. 7.6. The gas of interest is placed in a long cylindrical vessel. Inside the gas there is a concentric tube. A fan located in the lower part to the vessel creates a small vertical circulation of the gas.

If the gas pressure is comparatively low, the gas can be regarded to a good approximation as ideal since the enthalpy does not depend on the pressure in an ideal gas, so that

$$\left(\tfrac{\partial i}{\partial p}\right)_T = 0 \tag{7.46}$$

and, consequently, from the obvious relation

$$di = \left(\tfrac{\partial i}{\partial T}\right)_p dT + \left(\tfrac{\partial i}{\partial p}\right)_T dp, \tag{7.47}$$

which, using (1.34), is written in the form

$$di = c_p dT + \left(\tfrac{\partial i}{\partial p}\right)_T dp, \tag{7.47a}$$

we find for the case of interest

$$di = c_p dT. \tag{7.48}$$

We substitute this quantity into Eq. (7.44a) and find

$$c_p dT = -dh, \tag{7.49}$$

i.e., when an ideal gas rises (dh > 0) on adiabatic flow the temperature of the gas falls (dT < 0).

Integrating (7.49) between limit points 1 and 2 (Fig. 7.6), we find

$$-\int_{T_1}^{T_2} c_p dT = h_2 - h_1 \tag{7.50}$$

or, what is the same,

$$c_p^{av}(T_1 - T_2) = h_2 - h_1. \tag{7.51}$$

where c_p^{av} is the average heat capacity c_p in the temperature range from T_1 to T_2.[†] If the temperature difference $(T_1 - T_2)$ is not too large, the average heat capacity c_p^{av} nearly equals the true heat capacity c_p. We then find from (7.51 that

$$c_p = \frac{h_2 - h_1}{T_1 - T_2}. \tag{7.52}$$

It is evident from this that by measuring the temperature difference of the gas between two known heights h_1 and h_2 we can find its heat capacity c_p. Experimental apparatus for measuring the heat capacity of air constructed using this method has a height of 3 m. Consequently $(h_2 - h_1)$ amounts to 3 kg·m/kg: with 1 kcal = 426.94 kg·m, this value becomes 0.007 kcal/kg while the temperature difference is about 0.03°C. It should be noted that under experimental conditions of this kind good precision can only be obtained for those materials which have not too great a thermal conductivity, since for a high thermal conductivity heat transport along the flow becomes substantial and the adiabatic flow conditions are destroyed.

7.5.   Thermodynamics of the Atmosphere

Process which are rather similar to those treated in the preceding section occur in the atmosphere. As we know, with increasing height the temperature of the air in the atmosphere decreases. This drop in the temperature of atmospheric air with height is largely due to circulation of the air (warm air in the lower

---

[†]Remember that the average heat capacity c_p between temperatures T_1 and T_2 is given by the obvious relation

$$c_p^{av} = \frac{\int_{T_1}^{T_2} c_p dT}{T_2 - T_1}.$$

layers of the atmosphere is less dense than cold air in the upper layers; this is the reason for the natural circulation of the air).

Since the flow of heat due to vertical thermal conduction is quite small because of the comparatively low thermal conductivity of air, the circulation of air between the upper and lower layers of the atmosphere can be regarded as nearly adiabatic.

We will find the distribution of temperature and pressure with height in the atmosphere.

As we know from general thermodynamics, the relation between the pressure p and temperature T in an adiabatic process in an ideal gas takes the form

$$pT^{-\frac{k}{k-1}} = \text{const.} \tag{7.53}$$

Taking the logarithm and then differentiating this relation, we find

$$\frac{dp}{p} = \frac{k}{k-1}\frac{dT}{T}, \tag{7.54}$$

which is the differential equation for the adiabatic curves of an ideal gas.

As shown above, the differential equation for the pressure variation with height in a column of an ideal gas [Eq. (7.20)] is

$$\frac{dp}{p} = -\frac{dh}{RT}.$$

Replacing dp/p in this equation using Eq. (7.54), we find

$$\frac{dT}{dh} = -\frac{k-1}{kR}. \tag{7.55}$$

This equation determines the vertical temperature gradient in the atmosphere. As we see from this equation, since $k > 1$ we have $dT/dh < 0$, i.e., the temperature of the atmosphere falls with increasing height. Since k and R naturally do not change with height, the gradient dT/dh preserves the same value for all heights.[†]

[†]In reality this is not always true. One reason for that well-known disaster "smog" (pollution and stagnation of the air above cities), is inversion of the temperature gradient in the atmosphere. For more details see L. Batgan, "The Polluted Sky," Izdatel'stvo "Mir" (1969).

For air the gas constant R = 29.28 kcal/kg · deg while the adiabatic exponent is k = 1.41. Substituting these values into (7.55), we find for the atmosphere

$$\frac{dT}{dh} = -0.00993 \text{ deg/m}.$$

In other words, with each kilometer rise the temperature of the air falls linearly by 9.93°C. In reality the vertical temperature gradient in the atmosphere has a value lower than −9.9°C. This is because the water vapor in the air condenses when a hot mass of air rises into the cooler upper layers of the atmosphere and releases the heat of condensation during its phase transformation, retarding the cooling of the air. As shown by measurements, the actual vertical temperature gradient in the atmosphere is −6.5°C (this gradient holds practically constant up to a height of 11 km). Since

$$\frac{dT}{dh} = -\frac{k-1}{kR} = \text{const},$$

it is clear that the relation between air temperature and height is

$$T = T_0 - \frac{k-1}{kR} h, \tag{7.56}$$

where T_0 is the temperature of the air at the surface of the earth (h = 0).

Substituting this expression for T(h) into Eq. (7.20), we find that

$$\frac{dp}{p} = -\frac{dh}{RT_0 - \frac{k-1}{k} h}. \tag{7.57}$$

Integrating this relation between h = 0 and h (this corresponds to pressures from p_0 to p), we find

$$\ln \frac{p}{p_0} = \frac{k}{k-1} \ln \frac{RT_0 - \frac{k-1}{k} h}{RT_0} \tag{7.58}$$

or, what is the same,

$$p = p_0 \left(1 - \frac{k-1}{kRT_0} h\right)^{\frac{k}{k-1}} \tag{7.59}$$

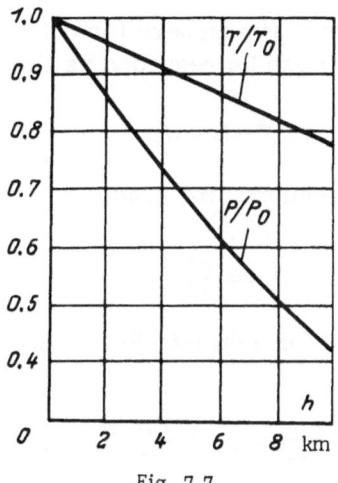

Fig. 7.7

This is the form of the relation between atmospheric pressure and height. Taking (7.56) into account this relation can be transformed to

$$p = p_0 \left(\frac{T}{T_0}\right)^{\frac{k}{k-1}},$$

(7.60)

i.e., as we would expect we have found the usual equation for the adiabatic curves of an ideal gas.

The nature of the relation between the temperature and pressure of the atmospheric air and the height is shown in Fig. 7.7.

In conclusion, we discuss the so-called standard atmosphere. As is well known, in technology the relation between atmospheric pressure and height is often used to determine the height above sea level from the results of pressure measurements using barometric height meters (altimeters). These devices usually have two scales, calibrated in pressure units (torr or meters of water) and in meters above sea level. In this regard the question arises: how is the height scale to be calibrated, i.e., how are we to establish the relation h(p) for the atmosphere?

For an ideal-gas atmosphere, in agreement with (7.21), we have

$$\ln \frac{p(h)}{p_0} = -\frac{1}{R} \int_0^h \frac{dh}{T}.$$

As shown above, for a real atmosphere for not too large height changes the relation T(h) can be regarded as linear:

$$T = T_0 + \tau h, \qquad (7.61)$$

where τ is the vertical temperature gradient

$$\tau = \frac{dT}{dh}.$$

Substituting this relation for T(h) into Eq. (7.21) and integrating, we find

$$\ln \frac{p(h)}{p_0} = \frac{1}{R\tau} \ln \frac{T_0 - \tau h}{T_0}, \qquad (7.62)$$

whence

$$h = \frac{T_0}{\tau} \left[1 - \left(\frac{p}{p_0} \right)^{R\tau} \right]. \qquad (7.63)$$

Knowing the vertical temperature gradient τ we can use this semiempirical barometric formula to determine the height h from the measured pressure p.

The barometric formula (7.63) is used to calculate the so-called standard atmosphere widely used in meteorology. The standard atmosphere is defined as the pressure distribution with height given by the barometric formula (7.63) assuming that the pressure at sea level with T_0 = 15°C is 760 torr and τ = -6.5 deg/km for h < 11 km (consequently at this height T =-56.5°C) and τ = 0 for h between 11 and 25 km. Under these assumptions we find from (7.63) that

$$h = 44{,}308 \left[1 - \left(\frac{p}{p_0} \right)^{0.19028} \right] \text{ for } h \leqslant 11 \,\text{km} \qquad (7.64)$$

and

$$h = 11{,}000 + 6\,340 \ln \frac{p_{11}}{p} \text{ for } 11\,\text{km} < h < 25\,\text{km} \qquad (7.65)$$

where p_{11} is the pressure at a height of 11 km and h is given in meters. The height scales of barometric altimeters are also calibrated using this standard atmosphere.

Naturally the question arises: why is the special height of 11 km the boundary between the applicability of Eq. (7.64) and Eq. (7.65)?

The answer to this question requires some information about the earth's atmosphere.

The lower layer of the atmosphere is called the troposphere. In the troposphere there are downgoing and upgoing air flows whose origins were discussed above.

At a height of about 11 km (near the earth's poles this level drops to 6 km and above the equator it rises to 18 km) the troposphere gives way to the stratosphere — an isothermal region which extends about 10 to 14 km in height.

Above the stratosphere lies the so-called upper layer of the atmosphere.

In entering the upper layer of the atmosphere the temperature increases. At a height of 50 km the temperature reaches 0°C. This increase in the temperature is due to an increase in the ozone content in the air at these heights; ozone is an intense absorber of solar radiation, this causes some heating of this layer of the atmosphere. With a further increase in height the ozone concentration in the air decreases and the air temperature again begins to drop. At a height of about 80 km the temperature has fallen to −88°C.

Above 80 km the air becomes ever more dissociated with increasing height and eventually also ionized. Due to the dissociation and ionization the optical transparency of the air decreases and it absorbs solar radiation, which leads to an increase in the air temperature at heights of more than 80 km. This increase in the temperature turns out to be quite substantial: as shown by measurements made with satellites at 500 km the temperature is 1680°C.

At the present time the standard atmosphere has been calculated up to a height of 200 km. These calculations are done taking all of the above-described physical processes which occur in the upper layer of the atmosphere into account.

CHAPTER 8

Liquids in the Weightless State

8.1. Features of the Behavior of Two-Phase Systems in the Weightless State

The behavior of liquids in the weightless state is of great interest primarily for the technology of space flight. Many important regularities in the behavior of liquids under weightless conditions can be found from thermodynamics. Below we will treat liquids in equilibrium with their vapor.

In Section 6.4 it was shown that a liquid under weightless conditions surrounded on all sides by its vapor will assume a spherical shape (the shape corresponding to least free energy of the system). Here we treat a more complicated question. Under weightless conditions what behavior is to be expected for a liquid contained in a vessel? The complexity arises because in this case we must consider the interaction of the liquid with the solid walls of the vessel (the contact angle θ, etc.).

In fact, the solution of the problem will be different for vessels of different shapes. Here we consider the behavior of a liquid in a vessel having the simplest geometrical shape — a sphere. Under ordinary conditions in the presence of gravitation liquid is found in the lower part of a vessel and vapor or gas in the upper part. The position of a wetting liquid in a vessel under these conditions is shown in Fig. 8.1a.

We might expect that with the advent of weightlessness the

193

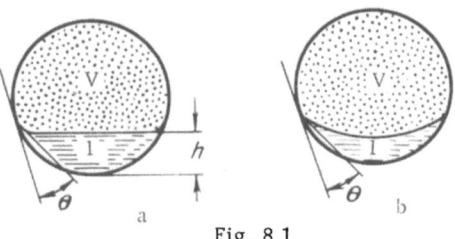

Fig. 8.1

shape of the surface separating the liquid and vapor would change,
since there is a change in the relationship of the forces which act
on the system (the weight force vanishes). Here two cases are
possible in principle†: (1) both the liquid and its vapor make contact
with the solid walls of the vessel, and (2) one of the phases is completely out of contact with the walls of the vessel and lies within
the other phase.

    In the first case the surface separating the liquid and vapor
phases will join the solid walls of the vessel at the contact angle θ
(the contact angle θ will be the same under weightless conditions
as in the presence of gravity‡). On the other hand, at the liquid–
vapor boundary the surface tension force creates a pressure
difference between the coexisting phases which is given by Laplace's
equation (6.67):

$$p_1 - p_2 = \sigma \left(\frac{1}{\rho_I} + \frac{1}{\rho_{II}} \right).$$

    Naturally in the absence of external forces the surface
separating the two phases has the shape for which the pressure
is the same at all points of the surface, i.e., the sum $(1/\rho_I + 1/\rho_{II})$
is the same. As shown by geometrical analysis, the unique surface
which satisfies the condition

$$\frac{1}{\rho_I} + \frac{1}{\rho_{II}} = \text{const}$$

(8.1)

is a spherical surface. Thus if both phases — liquid and vapor —

---

†We do not consider the case where one of the phases divides into several bubbles or
drops. Such an arrangement can really only occur if there is still some external
force field (vibrations, for example). This arrangement is clearly not favorable thermodynamically, since there is an increase in the total area of the surface separating
the phases and thus the free energy of the system increases.

‡This is valid not too close to the critical point, near which the thickness of the transition layer between liquid and gas increases, the compressibility increases, and the
weight force thus makes a substantial contribution to the surface energy.

 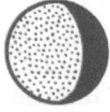 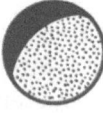

Fig. 8.2

are in contact with the vessel walls under weightless conditions, the wetting liquid lies as shown in Fig. 8.1b; the surface of separation assumes a spherical shape. It is important to note here that since there is no difference between "up" and "down" under weightless conditions, the liquid can make contact with the walls at any part of the vessel (see Fig. 8.2; all of the positions of the liquid in the vessel shown in this figure are equally valid thermodynamically).

In the second case, where one of the phases makes no contact with the walls of the vessel, this phase, as noted above, will take a spherical shape (recall that not only can the liquid gather in a sphere surrounded on all sides by vapor, but the vapor volume can be a spherical inclusion in the liquid). It is easy to see that if the contact angle is $0°$ or $180°$ the first case coincides with the second and is the only one possible under the given conditions; with $\theta = 180°$ the liquid is surrounded by the vapor, with $\theta = 0°$ the vapor collects in the center of the liquid, which covers all of the inner surface of the spherical vessel. For all other values of θ, i.e., $0 < \theta < 180°$, one experimentally observes both the arrangement corresponding to the first case (the liquid–vapor interface intersecting the walls of the vessel) and the arrangement corresponding to the second case (one phase within the other) although, as shown by the analysis given above, in this case the liquid–vapor phase separation surface should remain in contact with the walls of the vessel. For example, experiments are known in which under weightless conditions mercury separates from the glass surface of a vessel (for mercury–vapor–glass we have $\theta \approx 130°$).

It is of interest to give a more detailed analysis of the phenomena. In what follows we consider only the case where $0 < \theta < 180°$.

8.2. Possibility That One of the Phases May Lose Contact with the Walls of the Vessel

We consider a spherical vessel of radius R initially filled (under normal gravity) with a liquid to height h. Above the surface

of the liquid is its vapor. The temperature T of the system will
be assumed constant. Since the radius of the vessel does not change
it is evident that the system of interest is an isochoric—isothermal
system (V_{syst} = const, T_{syst} = const).

   As is well known, under ordinary conditions in the presence
of gravitation the surface of the liquid in a vessel will be horizontal
(if we neglect a certain curvature of the liquid surface at the vessel walls due to wetting effects; this neglect is justified for vessels of sufficiently large radius). The conditions for a horizontal
surface follow from elementary thermodynamic consideration. In
agreement with (2.24) the conditions for equilibrium in an isochoric-
isothermal system are written in the following way:

$$dF + dL^* \leqslant 0.$$

We recall that for a system in a gravitational field [Eq. (7.5)]

$$dL^* = Gdh.$$

In view of this relation the equilibrium condition (2.24) can be
written as

$$dF + Gdh \leqslant 0, \qquad\qquad (8.2)$$

so that for systems in which G = const we have

$$d(F + Gh) \leqslant 0, \qquad\qquad (8.3)$$

where h is the height of the center of mass of the system. Thus
the system is in equilibrium if the sum (F + Gh) is a minimum.
As shown in Section 6.3, the relation for the free energy of the
mass of liquid (which is not attached to the walls of the vessel)
is [Eq. (6.39)]

$$F = fG + \sigma \mathfrak{S}.$$

   As can easily be shown, $\sigma \mathfrak{S}$ is negligible compared to Gh.
It is thus evident that with V_{syst} = const and T_{syst} = const the sum
(F + Gh) is a minimum when h has its minimum value. It is clear
from elementary geometrical considerations that the lowest position
of the center of gravity of the liquid occurs with the liquid surface
horizontal.

Thus in the presence of gravity a liquid in a spherical vessel takes the form of a spherical segment in the lower part of the vessel.

For weightless conditions the dL^* term in Eq. (2.24) is absent and the equilibrium state of the liquid will be determined by the minimum in the free energy of the liquid–vapor–walls system. It is evident that with the onset of weightlessness the initial (horizontal) position of the liquid in the vessel in general does not correspond to a minimum of the free energy and consequently is not the equilibrium position. Therefore in the isothermal system of interest there will be spontaneous processes leading to a decrease in the free energy (it is understood that the position of the center of mass of the system† changes). We will discuss this process. In the first place it should be stressed that, as shown above, when $0 < \theta < 180°$ the position where the spherical surface of the liquid intersects the walls of the vessel corresponds to the minimum free energy for the given system (F_{min}). Consequently for this case ($0 < \theta < 180°$) the position where one phase as a whole lies inside the other corresponds to a free energy value greater than F_{min} (but in fact less than the initial free energy of the system with a planar phase separation surface — otherwise the position of the separation surface would remain unchanged).‡ We denote the free energy of the system in the initial state where the surface separating the phases is horizontal by F_{init}, while the free energy of the system in the intermediate state where one of the phases lies wholly within the other is denoted by F_{int}. As we have noted,

$$F_{init} > F_{int} > F_{min} \qquad (8.4)$$

(see Fig. 8.3; in this figure a, b, and c are, respectively, the initial state, the intermediate state, and the equilibrium state).

---

†It is evident that for weightless conditions instead of the idea of "center of gravity" we must use the idea of "center of mass." In general we should also take account of the effect of the intrinsic gravitational field created by the mass of the liquid of interest. However, this effect becomes significant only for liquid masses of thousands of tons.

‡The relation $F_{int} > F_{min}$ will be derived in detail.

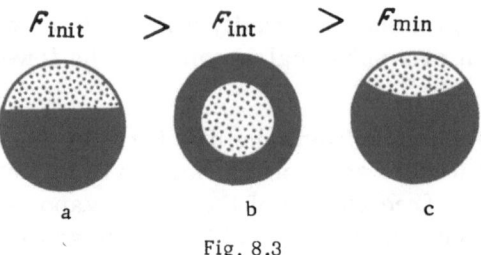

Fig. 8.3

After the onset of the weightless state a process begins in the system which is accompanied by a change in the position of the liquid–vapor phase separation surface and consequently a change in the contact area between phases. We assume that during this process the system goes into an intermediate state with free energy F_{int} (one phase within the other).

We denote the changes in the contact area of the liquid with the solid walls of the vessel, the vapor with the solid walls of the vessel, and the liquid with the vapor by $\Delta\mathfrak{S}_{1-s}$, $\Delta\mathfrak{S}_{v-s}$, and $\Delta\mathfrak{S}_{1-v}$. It is clear that

$$\Delta\mathfrak{S}_{1-s} = -\Delta\mathfrak{S}_{v-s} . \tag{8.5}$$

It is further clear that the change (drop) in the free energy of this system in going from the initial to the intermediate state ($\Delta F = F_{init} - F_{int}$) can be written, in agreement with Eq. (6.17), as

$$\Delta F = \sigma_{1-s}\, \Delta\mathfrak{S}_{1-s} + \sigma_{v-s}\Delta\mathfrak{S}_{v-s} + \sigma\Delta\mathfrak{S}_{1-v} , \tag{8.6}$$

where σ_{1-s}, σ_{v-s}, and σ are the surface tension coefficients at the solid–liquid boundary, the solid–vapor boundary, and the liquid–vapor boundary. Using (8.5), we can write this relation in the form

$$\Delta F = (\sigma_{1-s} - \sigma_{v-s})\Delta\mathfrak{S}_{1-s} + \sigma\Delta\mathfrak{S}_{1-v} . \tag{8.7}$$

Since in agreement with Eq. (6.91)

$$\frac{\sigma_{v-s} - \sigma_{1-s}}{\sigma} = \cos\theta,$$

where θ is the contact angle of the system, we find from (8.7) that

$$\Delta F = \sigma(\Delta\mathfrak{S}_{1-v} - \Delta\mathfrak{S}_{1-s} \cos\theta). \tag{8.8}$$

We inquire about cases in which one of the phases forms a sphere inside the other phase. When is this possible in principle, even if it is only one of the intermediate states in the processes occurring after the vanishing of the gravitational force?

We recall that spontaneous processes in an isochoric–isothermal system are subject to the condition (2.14)

$$\Delta F \leqslant 0$$

(the equality sign corresponds to the equilibrium state of the system). In view of (8.8) we have

$$\sigma(\Delta \mathfrak{S}_{1-v} - \Delta \mathfrak{S}_{1-s} \cos\theta) \leqslant 0$$

or

$$\cos\theta \leqslant \frac{\Delta \mathfrak{S}_{1-v}}{\Delta \mathfrak{S}_{1-s}}. \tag{8.9}$$

It should be understood that since the shape of the vessel is fixed (a sphere), $\Delta \mathfrak{S}_{1-v}$ and $\Delta \mathfrak{S}_{1-s}$ are uniquely related to each other if we know what fraction of the volume of the vessel is occupied by liquid, i.e., if we know the relation h/R, where h is the height of the spherical segment occupied by liquid under normal gravity and R is the radius of the vessel.

On the other hand, the contact angle θ is a constant for a given liquid–solid pair.

As we see from (8.9), the case considered where the system transforms into an intermediate state can occur only for values of h/R for which $\Delta \mathfrak{S}_{1-v} / \Delta \mathfrak{S}_{1-s}$ is not less than cos θ. Here the equality sign in Eq. (8.9) corresponds to the case where $\Delta F = 0$, i.e.,

$$F_{\text{init}} = F_{\text{int}} \tag{8.10}$$

and consequently a transition from the initial to the intermediate state is always possible (there is no difference between these states from the viewpoint of the free energy). A transition is also possible when

$$\cos\theta < \Delta \mathfrak{S}_{1-v} / \Delta \mathfrak{S}_{1-s}. \tag{8.11}$$

If the filling (h/R) of the vessel is such that

$$\cos \theta > \Delta \mathcal{E}_{1-v} / \Delta \mathcal{E}_{1-s} ,$$ (8.12)

then no transition from the initial to the intermediate state is possible since in this case

$$F_{int} > F_{init}$$ (8.13)

Assuming different values for (h/R) we can calculate $\Delta \mathcal{E}_{1-v} / \Delta \mathcal{E}_{1-s}$ and then calculate the value of h/R for which Eq. (8.9) is satisfied for any given value of θ. The results of such a calculation are shown in Fig. 8.4, in which the coordinates $\theta = f(h/R)$ are used to show curves separating regions A, B, and C. The values of θ lying at the boundaries of these regions will be called the limiting values of the contact angles. If the parameters of the system are given by a point lying in region A, it is thermodynamically possible to have a transition under conditions of complete weightlessness from the initial state into a state in which the gas forms a sphere within the liquid. In region B it is possible for the liquid to separate from the walls of the vessel and appear as a sphere within a gaseous "pocket." If the parameters of the system are given by a point lying in region C, separation of one phase from the walls of the vessels under weightless conditions is not possible in principle due to the effect of surface forces and cannot be achieved in a correctly set-up experiment. Otherwise we would have a spontaneous process with an increase of the free energy.

Since the gas—liquid separation surface area is smaller when one of the phases lies as a sphere within the other phase than in all other cases where one phase separates from the walls of the vessel, the free energy is lowest in this case. The calculations which have been carried out give a general solution of the question of whether it is possible in principle to have separation of one of the phases from the walls of a spherical vessel.

The conclusion that there exist limiting values of the wetting angle which bound regions where separation of one phase from the walls of the vessel is possible is of definite independent interest. Indeed, for example, water froms a contact angle of the order of 45° with many metals, i.e., it is a wetting liquid. As shown in Fig. 8.4, when such a metal vessel is filled halfway with water [(h/R) = 1], it is generally not possible for it to completely separate

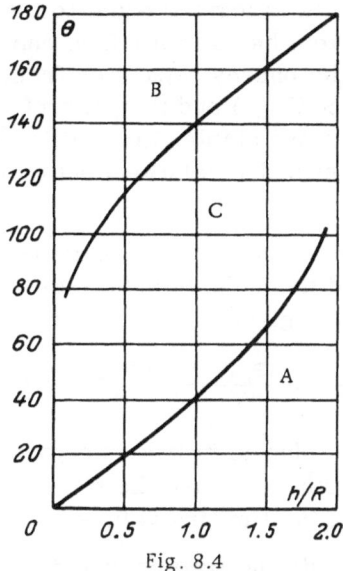

Fig. 8.4

from the walls of the vessel. With h/R > 1.1 such a transition
becomes possible in principle. On the other hand, if mercury is
in a glass vessel it can separate from the walls in principle only
with h/R < 0.8 (the contact angle depends substantially on the
cleanliness of the surface). (We have taken θ = 130° for mercury.)
It should be noted in passing that experimentally in one case mercury separates from the walls and in another case it does not.

It is of interest to note that for large initial fillings (h/R >
1.82) even with a nonwetting liquid the gas can exist as a sphere
inside the liquid. This happens because for such values of h/R
if the gas forms a sphere within the liquid, the liquid–gas surface
area will be less than in the initial position. The drop in free
energy which is involved may overcome the increase in the free
energy due to the increase in the fraction of surface area of the
solid in contact with the nonwetting liquid. Correspondingly,
for small values of h/R in principle the liquid can lie within a
gaseous "pocket" even with θ < 90°.

8.3. Stable Equilibrium States of a Two-Phase System

We will now show that in a state in which the liquid–vapor
separation surface is spherical and intersects the walls of the

vessel such that an angle is formed equal to the contact angle θ of the system, the system has a lower free energy than in the state in which one of the phases separates from the walls of the vessel (i.e., that $F_{int} > F_{min}$) and that, therefore, of these two states, the first is the more stable state. It is of interest to treat the corresponding quantitative relations between free energy values in these cases.

The change in the surface free energy relative to the initial state can be calculated from Eq. (8.8) for a given value of θ. This equation is conveniently given in dimensionless form as

$$\frac{\Delta F}{\pi R^2 \sigma} = \frac{\Delta \mathfrak{S}_{1-v}}{\pi R^2} - \frac{\Delta \mathfrak{S}_{1-s}}{\pi R^2} \cos \theta. \qquad (8.14)$$

In calculating the change in free energy when one of the phases lies within the other, we use the same values of $\Delta \mathfrak{S}_{1-v}$ and $\Delta \mathfrak{S}_{1-s}$ as in calculating the limiting values of the contact angles.

In order to calculate the free energy difference between the initial state with a planar surface between phases and the state of stable equilibrium under weightless conditions we need to calculate the geometrical properties of the sphere which is formed when the liquid (vapor) intersects the spherical surface of the vessel at a given angle θ and divides the vessel into two parts whose relative sizes are determined by the amount of liquid poured into the vessel. The solution of this problem, although involving quite cumbersome calculations, presents no difficulties of principle.

Figure 8.5 shows the changes in the surface of contact between the two media in going from the initial state to the equilibrium state under weightless conditions and also the position of the liquid under weightless conditions as a function of h/R. Here the wetting angle is taken to be 45°. Curve 1 refers to the liquid–gas separation surface and curve 2 to the liquid–solid separation surface.

We can see that for h/R < 0.293 at equilibrium under weightless conditions the liquid–gas separation surface is convex on the gas side. The area of this surface is less than at the initial position. This decrease in surface area correspondingly decreases the free energy. In fact, there is also a decrease in the surface area of solid–liquid contact and a corresponding increase in the

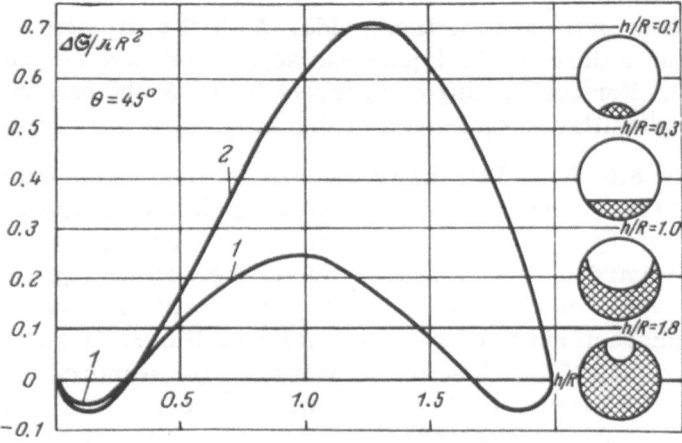

Fig. 8.5

solid–gas surface area. Since for θ < 90° the free energy per
unit area of solid–gas surface is greater than that for the solid-
liquid surface, the latter process increases the free energy. However, as shown by calculations, the free energy as a whole decreases.

As h/R varies from 0 to 0.293 the radius of the liquid–gas
separation surface under equilibrium conditions with no gravity
increases from 0 to ∞. With h/R ≈ 0.293 in the presence of
gravity a planar liquid–gas separation surface forms, intersecting
the spherical walls of the vessel at 45°, i.e., at an angle equal
to the assumed value of the wetting angle. In this case the surface
of separation remains planar whether gravitational forces act or
not.

With 0.293 < h/R < 1.67 the area of the liquid–gas surface
increases in going to weightless conditions, but there is also an
increase in the amount of solid surface occupied by liquid. The
free energy of the system as a whole decreases.

Finally, with h/R > 1.67 the free energy decreases due to
a decrease in the liquid–gas surface area and also due to an increase
in the portion of the solid surface in contact with the liquid.

Over the whole region from h/R = 0.293 to h/R = 2 in
equilibrium under weightless conditions the liquid–gas separation

surface is convex toward the liquid side. As h/R varies from
0.293 to 2 the radius of the liquid—gas separation surface decreases
from ∞ to 0. For contact angles different from 45° the pattern is
qualitatively similar to that given in Fig. 8.5.

Figure 8.6 shows how the free energy of the system changes
under weightless conditions for the following characteristic values
of the wetting angle: 15°, 45°, 135°, and 165°. A contact angle of
15° can be regarded as an average value of the wetting angle between
glass and water (for glass which is not very clean). An angle of
$\theta = 45°$ is characteristic of water wetting metallic surfaces. A
contact angle of 135° is close to the wetting angle of mercury with
glass and is the wetting angle for the water—zinc stearate system.
The value $\theta = 165°$ corresponds to poor wetting of the surface.

Curve 1 shows the change in the free energy in going from
the initial state to a state in which one of the phases is completely
separated from the walls of the vessel and takes the form of a
sphere within the other phase. Part of these lines lie above the line
$\Delta F/\pi R^2 \sigma = 0$, which means that a transition to this state is not
possible. The intersection point of curve 1 with the line $\Delta F/\pi R^2 \sigma = 0$
agrees with the intersection point of the corresponding ordinates
in Fig. 8.4 with the limiting curves.

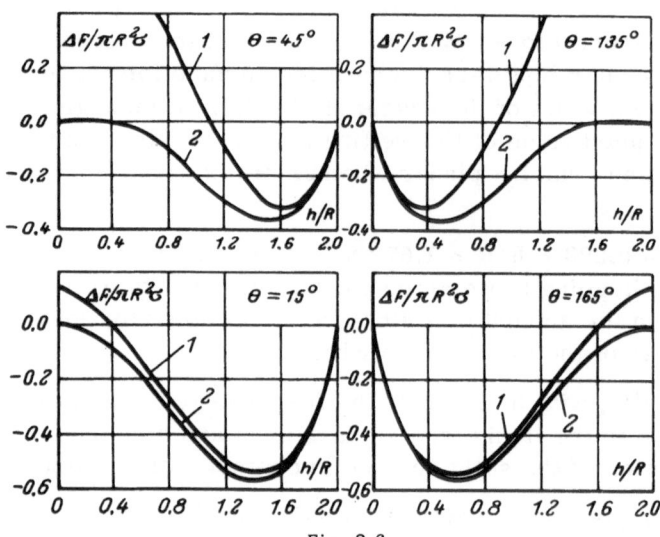

Fig. 8.6

Curve 2 (Fig. 8.6) shows the change in free energy in going from the initial state to the equilibrium state under weightless conditions. Naturally it lies below line 1 over its entire length. Beginning at a certain value of h/R the curves nearly coincide. This is especially prominent for values of θ close to 0 or 180°. In these regions the differences in the free energy between the equilibrium state and the state in which one of the phases separates from the wall are very small and a random perturbation can carry the system from one state to the other.

It is useful to note that curve 2 for the change of free energy has not one but two minima lying on different sides of $\Delta F/\pi R^2 \sigma = 0$ (the point $\Delta F = 0$ corresponds to an amount of liquid in the vessel for which the planar phase separation surface would intersect the spherical walls of the vessel at an angle equal to the wetting angle in the presence of gravity. For values of θ which differ substantially from 90° one of the minima is barely perceptible.

The thermodynamic analysis given above allows us to give a qualitative treatment of the transformation process from the initial state with a planar phase separation surface to the equilibrium state under weightless conditions.

With the onset of weightlessness an irreversible process begins which decreases the free energy of the system. The position of the liquid oscillates about the equilibrium state. In viscous

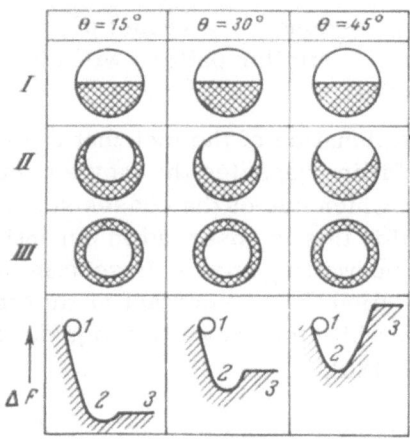

Fig. 8.7

liquids this process is damped. It should be mentioned that for
initial states of the system represented by points lying in regions
A and B of Fig. 8.4 the kinetic energy which the liquid possesses when
it passes through the state with minimum surface free energy can
turn out to be sufficient that the layer of one of the phases completely
closes and separates the other phase from the wall.

Here we can give a simple mechanical analogy. In Fig. 8.7 the initial state
of the system is shown as the position of the heavy sphere on the curved surface at
point 1. Here the potential energy of the sphere is identified with the excess free
energy of the system relative to the equilibrium position. Here all of the vertical
dimensions are given in a scale corresponding to the actual case of a spherical vessel
filled halfway with liquid, with wetting angles equal to those given in Fig. 8.7. All
horizontal dimensions are arbitrary. Position 2 is a state of stable equilibrium. Position 3 corresponds to separation of the gas phase from the walls. The presence of a
horizontal area indicates that with regard to energy it makes no difference in what
part of the vessel the gaseous sphere which is detached from the walls is located.

In complete agreement with the analysis given above, it is not possible for the
sphere to fall in position 3 with a wetting angle of 45°. Whether the system goes
into position 3 with wetting angles of 30° and 15° depends only on the amount of
friction. For sufficiently low liquid viscosities separation of the gas phase from the
walls is inevitable in this case.

In treating the mechanical model we see that in principle the sphere can be
found on the horizontal part 3 for long times. However, a small perturbation is sufficient to put it into the equilibrium state 2. It should be noted that a strong perturbation corresponding to the characteristic energy required to upset the real system can
cause the sphere to go into a state close to the initial state and again fall into state 3.

The analysis has been done for a vessel of spherical shape;
however, it is evident that a similar pattern will also be observed
for vessels of other shapes.

We thus find an explanation of the fact that in brief experiments done largely in falling capsules the movie camera shows
a state of the system in which one of the phases separates from the
walls. For wetting liquids this is observed in essentially all such
experiments, while for nonwetting ones it is seen in only some.
Here we should bear in mind that the contact angle for a wetting liquid
is often very close to 0, while for nonwetting liquids the contact
angle is always close to 180°.

CHAPTER 9

Radiation

9.1. Radiation in a Cavity as a Thermodynamic System

Thermodynamic analysis of the equilibrium of electromagnetic radiation leads to important results. We consider a closed cavity containing a vacuum. Such a cavity can be represented, for example, as a cylinder with a movable piston. The inside of the cavity is a perfect blackbody at a temperature T. This perfect blackbody radiates into the cavity; consequently the cavity has a certain energy. We assume that the sides of the cylinder and the piston are made from completely reflecting material (this assumption is made to simplify the calculations — it permits us to neglect heat exchange with the walls of the cavity).

As we know from physics, the radiation from a perfect blackbody is equilibrium radiation. If the radiation is in equilibrium with a body at a temperature T then we can say that the radiation which fills the cavity is also at temperature T.

The idea of radiation temperature was first introduced by B. B. Golitsyn (1893). This idea allows us to use thermodynamic methods to treat the radiation. It should be stressed that the idea of a radiation temperature applies only to equilibrium radiation (therefore we will consider only radiation from a perfect blackbody).

If the dimensions of the cavity are sufficiently large relative to the radiation wavelength then the radiation in the cavity can be regarded as homogeneous, i.e., the same at all points.

Thus in view of the remarks above it is evident that the radiation in the cavity is a thermodynamic system, to which one can supply or remove heat (with the aid of a perfect blackbody within the cavity) and which can be expanded (by moving the piston).

We now introduce the idea of r a d i a t i o n   d e n s i t y. By radiation density (u_V) we mean the amount of radiation energy per unit volume of the cavity (in other words, the volume-specific energy

$$u_V = \frac{U}{V},\qquad(9.1)$$

where U is the total energy of the radiation in the cavity and V is the cavity volume).

As we know from general physics, the radiation density is different for different wavelengths. We will consider the integrated radiation density

$$u_V = \int\limits_0^\infty u_{\lambda_V} d\lambda,$$

$$(9.2)$$

where u_{λ_V} is the density of radiation in the wavelength interval from λ to $\lambda + d\lambda$.

It is known that the radiation density depends on the temperature; the greater the temperature of the body the more energy it radiates. However, u_{λ_V} depends not only on the temperature but also on the wavelength λ. As for the integrated radiation density, it is clear from (9.2) that u_V depends only on the temperature.

It is shown in electrodynamics that i s o t r o p i c   e l e c t r o - m a g n e t i c   r a d i a t i o n   e x e r t s   p r e s s u r e   a t   a   s u r f a c e w h i c h   r e f l e c t s   o r   a b s o r b s   t h i s   r a d i a t i o n. The existence of light pressure was established experimentally by P. N. Lebedev in 1901.   In agreement with the laws of electrodynamics, the radiation pressure is uniquely related to the radiation density as follows:

$$p = \frac{u_V}{3}.$$

$$(9.3)$$

If radiation exerts a pressure on the walls of the cavity in which it is enclosed and this opposes a decrease in the volume of the cavity, then the radiation will do work against external pressure forces when the piston is moved.

It is not hard to see a definite (but completely superficial) analogy between equilibrium radiation and a gas. Both the radiation and the gas have pressure, in both cases the energy depends on the temperature, etc. Often the radiation in a cavity is referred to as a "photon gas" to stress this analogy. We recall that in quantum theory the radiation is thought of as a set of light "particles" — photons.

9.2. The Equations of State for a Photon Gas

It is clear from the discussion above that the usual thermodynamic rules apply to radiation in a cavity.

The combined equation for the first and second laws of thermodynamics is written for this case, just as for any other system, using the ability to do work of expansion against the external pressure forces:

$$TdS = dU + pdV. \tag{9.4}$$

Here, in agreement with (9.1),

$$U = u_V V, \tag{9.5}$$

where u_V is the radiation density (the volume-specific energy).

As noted above, u_V is a function of only the temperature and consequently is independent of the volume.[†]

---

†Here we make an analogy with an ideal gas, whose internal energy does not depend on the volume. However, there is no similarity. In an ideal gas the total internal energy U does not depend on the volume V of the gas (or the w e i g h t -specific internal energy u on the specific volume v):

$$\left(\frac{\partial U}{\partial V}\right)_T = \left(\frac{\partial u}{\partial v}\right)_T = 0.$$

As for the v o l u m e -specific internal energy u_V, since $u_V = u/v$ for an ideal gas, u_V depends on v while $u \neq f(v)$. For a photon gas u_V does not depend on v since $(\partial U/\partial V)_T \neq 0$.

In view of this we find from (9.5) that

$$\left(\frac{\partial U}{\partial V}\right)_T = u_V.$$

(9.6)

As we know from thermodynamics[†]

$$\left(\frac{\partial U}{\partial V}\right)_T = T\left(\frac{\partial p}{\partial T}\right)_V - p.$$

(9.7)

Substituting $(\partial U/\partial V)_T$ from (9.6) into this equation and p from (9.3), we find

$$u_V = \frac{T}{3}\frac{du_V}{dT} - \frac{u_V}{3}$$

(9.8)

(since u_V depends only on the temperature, the derivative of u_V with respect to T is a total derivative).

Solving this differential equation, we find

$$u_V = aT^4,$$

(9.9)

where a is a constant.

Equation (9.9), which gives the radiation density of a perfect blackbody as a function of the temperature, is called the Stefan–Boltzmann law. This law was experimentally established by J. Stefan in 1879 and derived by L. Boltzmann on the basis of thermodynamics in 1884.

The constant which enters Eq. (9.9) is

$$a = 1.82 \times 10^{-22} \text{ kcal/deg}^4 \cdot \text{cm}^3 = 7.64 \times 10^{-15} \text{ erg/deg}^4 \cdot \text{cm}^3 = 5.69 \times 10^{-8} \text{ W/cm}^2 \cdot \text{deg}^4$$

It follows from (9.9) that the energy of the photon gas vanishes at 0°K (since $u_V = 0$ at 0°K).

[†]This relation is not hard to obtain from Eq. (9.4) written in the form

$$dU = TdS - pdV.$$

From this equation it is clear that $\left(\frac{\partial U}{\partial V}\right)_T = T\left(\frac{\partial S}{\partial V}\right)_T - p,$ whence, taking (2.151a) into account, we find (9.7).

The total energy of the radiation in a volume V, in agreement with (9.5), is

$$U = aT^4 V. \tag{9.10}$$

In view of (9.9) we find from (9.3) the following relation for the radiation pressure as a function of temperature:

$$p = \frac{a}{3} T^4. \tag{9.11}$$

This relation can be regarded as the thermal equation of state for a photon gas, while the equation for the Stefan–Boltzmann law (9.9) can be regarded as the caloric equation of state for the photon gas.

It should be stressed that one characteristic feature of the photon gas is, as we see from (9.11), that the radiation pressure is uniquely related to the temperature; therefore an isothermal process in a photon gas is at the same time an isobaric process.

9.3. Entropy and Chemical Potential of the Photon Gas

It is evident that the expression for the entropy of a photon gas which occupies a certain volume V can be written in the form

$$S(V, S) = S(V = 0, T) + \int_0^V \left(\frac{\partial S}{\partial V}\right)_T dV, \tag{9.12}$$

where S(V, T) is the entropy of the photon gas at temperature T and volume V, while S(V = 0, T) is the entropy at the same temperature T but with V = 0. It is further evident that

$$S(V = 0, T) = 0.$$

Indeed, if S(V = 0, T) were not zero we would arrive at an absurdity: at V = 0 the system containing the photon gas does not rigorously exist, but its entropy differs from zero.

Thus

$$S(V, T) = \int_0^V \left(\frac{\partial S}{\partial V}\right)_T dV. \tag{9.12a}$$

We use the Maxwell equation (2.151a)

$$\left(\frac{\partial S}{\partial V}\right)_T = \left(\frac{\partial p}{\partial T}\right)_V.$$

The derivative $(\partial p/\partial T)_V$ will be calculated by differentiating Eq. (9.11):

$$\left(\frac{\partial p}{\partial T}\right)_V = \frac{4}{3} aT^3. \qquad (9.13)$$

In view of (9.13) we find from (9.12a) for the entropy of a photon gas in volume V

$$S(V,T) = \frac{4}{3} aT^3 V, \qquad (9.14)$$

whence the volume-specific entropy for the radiation is

$$s_V(T) = \frac{4}{3} aT^3. \qquad (9.15)$$

This quantity depends only on the temperature.

We should note yet another important fact. The isobaric–isothermal potential, which is determined by the general relation (2.16), which in view of (2.12) is written in the form

$$\Phi = U + pV - TS,$$

turns out to be zero for a photon gas. Indeed, taking account of the equations

$$U = aT^4 V, \qquad (9.10)$$

$$p = \frac{a}{3} T^4, \qquad (9.11)$$

and

$$S = \frac{4}{3} aT^3 V, \qquad (9.14)$$

we find $\Phi = 0$ from (2.16).

It is evident from this that the chemical potential Φ of a photon gas will also be zero. The definition of the chemical potential in the usual form (2.88), i.e., as the derivative of the potential with respect to the weight of the system $(\partial\Phi/\partial G)_{T,p}$ is quite mean-ingless for a photon gas. In this case the chemical potential should be defined as in chemical thermodynamics, as the derivative $(\partial\Phi/\partial N)_{T,p}$, where N is the number of particles (photons in the present case) in the system.[†]

It follows from the relation $\varphi = 0$ that the characteristic functions for a photon gas do not depend on the number of photons in the system. This is understandable since the photon gas is a special system: Photons are particles at the relativistic limit (the rest mass of a photon is zero) and therefore the number of photons in an isolated system is not constant, but rather the ther-modynamic parameters of the system are independent of the num-ber of particles [for example, as seen from Eq. (9.11), the pres-sure in a photon gas depends only on the temperature and is not related to the number of photons].

9.4. Thermodynamic Processes in a Photon Gas Heat Capacity

The volume-specific heat capacity C_V of a photon gas is defined by the obvious relation

$$C_v = \left(\frac{\partial S_V}{\partial T}\right)_V,$$

and is, using (9.15),

$$C_v = 4aT^3. \tag{9.16}$$

The heat capacity of the photon gas is quite small in absolute value; for example, at 3000°K we find from (9.16) that

$$C_v = 1.96 \times 10^{-11} \text{ kcal/cm}^3 \cdot \text{deg}$$

The heat capacity of radiation becomes comparable to the heat capacity of a monatomic ideal gas only at temperatures of the order of millions of degrees.

As for the isobaric heat capacity C_p of a photon gas, since as noted above for radiation an isobar is also an isotherm, and we

---

[†]It is clear that for ordinary systems these definitions are identical (within a constant factor):

$$\varphi = \left(\frac{\partial\Phi}{\partial G}\right)_{T,p} = \left(\frac{\partial N}{\partial G}\right)_{T,p}\left(\frac{\partial\Phi}{\partial N}\right)_{T,p};$$

so that clearly $(\partial N/\partial G)_{T,p}$ = const.

know that the heat capacity is infinite for an isothermal process, we find for the photon gas

$$C_p = \infty.$$

We will now formulate the relations which describe basic thermodynamic processes in a photon gas.

<u>Isothermal (Isobaric) Processes.</u> We calculate the amount of heat which must be supplied to a photon gas in order to expand it isothermally (isobarically) from volume V_1 to volume V_2.

The amount of heat supplied to the system (or removed from it) in an isothermal 1–2 process is given by the well-known relation

$$Q_{1-2} = T (S_2 - S_1). \tag{9.17}$$

In view of (9.14) we find for the change in the entropy in an isothermal process

$$S(T, V_2) - S(T, V_1) = \frac{4}{3} aT^3 (V_2 - V_1), \tag{9.18}$$

so that

$$Q_{1-2} = \frac{4}{3} aT^4 (V_2 - V_1) \tag{9.19}$$

or, what is the same,

$$Q_{1-2} = \frac{4}{3} (U_2 - U_1). \tag{9.20}$$

The work L_{1-2} done by a photon gas in isothermally expanding from volume V_1 to volume V_2 is defined by

$$L_{1-2} = \int_{V_1}^{V_2} p \, dV, \tag{9.21}$$

with the integral done along an isotherm.

Substituting p from (9.11) and assuming that an isothermal process is also isobaric in a photon gas [so that p can be removed from under the integral sign in Eq. (9.21)], we find the following relation for the work of expansion of a photon gas in an isothermal process:

$$L_{1-2} = \frac{aT^4}{3}(V_2 - V_1).$$ (9.22)

Adiabatic Process. To find the relations between V and T in an adiabatic (S = const) process we use the Maxwell equation (2.145a)

$$\left(\frac{\partial V}{\partial T}\right)_S = -\left(\frac{\partial S}{\partial p}\right)_V.$$

Since

$$\left(\frac{\partial S}{\partial p}\right)_V = \left(\frac{\partial S}{\partial T}\right)_V \left(\frac{\partial T}{\partial p}\right)_V,$$ (9.23)

in view of (9.13) and (9.14) we find

$$\left(\frac{\partial V}{\partial T}\right)_S = -\frac{3V}{T}.$$ (9.24)

Solving this differential equation by separation of variables and exponentiation, we find

$$VT^3 = \text{const.}$$ (9.25)

Replacing T in terms of p using Eq. (9.11), this equation becomes

$$pV^{4/3} = \text{const.}$$ (9.26)

This is the relation between p, V, and T for a photon gas in a reversible adiabatic process.

The structure of Eq. (9.26) recalls the Poisson equation for the adiabatic curves of an ideal gas (pV^k = const).

The work done by a photon gas in an adiabatic process between states 1 and 2 is determined using the general relation (9.21) given above; undoubtedly in the present case the integral in this relation should be calculated along an adiabatic curve. In agreement with (9.26) we can write for the adiabatic curves

$$p_1 V_1^{4/3} = pV^{4/3} , \qquad (9.27)$$

where p_1 and V_1 are the pressure and volume of the photon gas in the initial state 1 and p and V are the changing values of the pressure and volume. In view of this relation we find from (9.21) that

$$L_{1-2} = p_1 V_1^{4/3} \int_{V_1}^{V_2} \frac{dV}{V^{4/3}} , \qquad (9.28)$$

whence

$$L_{1-2} = 3p_1 V_1 \left[1 - \left(\frac{V_1}{V_2} \right)^{1/3} \right]. \qquad (9.29)$$

In an adiabatic process Q_{1-2} is naturally zero.

Isochoric Process. From the equation for the first law of thermodynamics for a system whose only form of work is work of expansion we find

$$dQ = dU + p \, dV, \qquad (9.30)$$

and it is clear that the amount of heat supplied to such a system (or removed from it) in a process with V = const is determined by

$$Q_{1-2} = U_2 - U_1. \qquad (9.31)$$

In view of (9.10) we find that

$$Q_{1-2} = aV (T_2^4 - T_1^4). \qquad (9.32)$$

Furthermore, since we see from (9.3) that

$$U = 3pV \qquad (9.33)$$

Eq. (9.31) can be represented in the form

$$Q_{1-2} = 3V (p_2 - p_1). \qquad (9.34)$$

Using Eq. (9.14) we find the following relation for the change in the entropy of a photon gas in an isochoric process:

$$S_2(V, T_2) - S_1(V, T_1) = \frac{4}{3} aV (T_2^3 - T_1^3). \qquad (9.35)$$

The work of expansion for a system in an isochoric process is naturally zero.

The Adiabatic Expansion of a Photon Gas in Vacuum. We consider an irreversible process in a photon gas —adiabatic expansion of the photon gas in vacuum without doing work.

We let the photon gas of interest be enclosed in a vessel separated by a moving mirror partition into two parts — one having volume V_1 and the other having volume V_{vac}. The vessel has mirror walls which prevent heat exchange with the external medium. Initially the photon gas occupies the part of the vessel having volume V_1 and the other part of the vessel (of volume V_{vac}) is empty. Then the partition is removed and the photon gas expands into volume V_{vac}. The expansion causes the pressure (and consequently also the temperature) of the photon gas to decrease, and its volume becomes equal to the whole volume of the vessel.

We will determine the changes in the temperature and entropy of the photon gas as a result of this process.

If the photon gas expands from volume V_1 to volume V_{vac} reversibly (not by moving the partition but, for example, by slow motion of a piston which does work against external forces), then the temperature change in this reversible adiabatic process can be determined using the previously obtained equation (9.25), from which it follows that

$$T_2 = T_1 \sqrt[3]{\frac{V_1}{V_2}}; \qquad (9.36)$$

and naturally the entropy of the photon gas does not change.

As we know, when the pressure in the system (here the pressure of the photon gas in volume V_1) does not equal the pressure in the surrounding medium (in our case in the volume V_{vac}), i.e., when the expansion process occurs in a nonequilibrium way, the differential work of expansion should be written in the form $p_c dV$, where p_c is the pressure of the surrounding medium. Consequently for systems whose only form of work is work of expansion the equation for the first law of thermodynamics (1.1) [using Eq. (1.17)] for the nonequilibrium expansion process is written in the form

$$dQ = dU + p_c dV. \tag{9.37}$$

Applying this to the process of adiabatic (dQ = 0) expansion of a photon gas in vacuum (p_c = 0 so that $p_c dV = 0$ — the photon gas does no work in expanding), we find

$$dU = 0 \tag{9.38}$$

and, consequently,

$$U = \text{const} \tag{9.39}$$

so that the internal energy does not change† in adiabatic expansion of a photon gas in vacuum.

It is further necessary to note the following. The differential equations of thermodynamics, as we know, are only applicable to reversible processes. It is thus evident that if it is possible to use these differential equations to calculate the changes in the temperature and entropy in an i r r e v e r s i b l e adiabatic expansion process of a photon gas from state 1 (volume V_1) to state 2 (volume V_2), we must first choose a scheme for a r e v e r s i b l e process in which the state of the photon gas goes from the same initial state 1 to the same final state 2. The change in the entropy ($S_2 - S_1$) and the change in the temperature ($T_2 - T_1$) will be calculated for this reversible process, and since the entropy and temperature are state functions it follows that ($S_2 - S_1$) and ($T_2 - T_1$) in this process will be the same as in an irreversible adiabatic expansion process.

It is evident that such a conventional reversible process could be, for example, reversible expansion of the photon gas with heat supplied, done in such a way that the internal energy U of the photon gas remains constant.

The change in the entropy of the photon gas due to this reversible process (equal to the change of entropy during irreversible adiabatic expansion into a vacuum) is given by the following obvious

---

†It is evident from the discussion scheme used here that this conclusion indeed holds not only for adiabatic expansion of a photon gas but also for adiabatic expansion (without doing work) of any gas.

relation:

$$S_2(U, V_2) - S_1(U, V_1) = \int_{V_1}^{V_2} \left(\frac{\partial S}{\partial V}\right)_U dV. \qquad (9.40)$$

From Eq. (1.28a) written (for the case where work of expansion is the only form of work) in the form

$$dU = T\, dS - p\, dV, \qquad (9.41)$$

it is obvious that

$$\left(\frac{\partial S}{\partial V}\right)_U = \frac{p}{T}. \qquad (9.42)$$

Substituting (9.42) into (9.40) and replacing p using Eq. (9.11) and assuming that, as we see from (9.10),

$$T = a^{-1/4} U^{1/4} V^{-1/4}, \qquad (9.43)$$

we obtain the following relation for the change in the entropy of a photon gas during adiabatic expansion into a vacuum without doing work:

$$S_2(U, V_2) - S_1(U, V_1) = \frac{4}{3} a^{1/4} U^{3/4} (V_2^{1/4} - V_1^{1/4}). \qquad (9.44)$$

It is thus evident that with $V_2 > V_1$ and $S_2 > S_1$ the entropy of the photon gas increases during the irreversible adiabatic expansion.

The change in the temperature of the photon gas during the process can be calculated using Eq. (9.43), from which it follows that

$$T_2(U, V_2) - T_1(U, V_1) = a^{-1/4} U^{1/4} (V_2^{-1/4} - V_1^{-1/4}). \qquad (9.45)$$

It is thus evident that with $V_2 > V_1$ and $T_2 < T_1$ the temperature of the photon gas can only decrease during irreversible adiabatic expansion.

Finally, the change in the pressure of the photon gas during this process is determined as follows. We write Eq. (9.3), using (9.1), in the form

$$p = \frac{U}{3V}, \qquad (9.46)$$

and find

$$p_2(U, V_2) - p_1(U, V_1) = \frac{U}{3}\left(\frac{1}{V_2} - \frac{1}{V_1}\right),$$ (9.47)

it being understood that with $V_2 > V_1$, $p_2 < p_1$.

The value U = const which enters Eqs. (9.40), (9.45), and (9.47) can be determined in terms of the parameters of the initial state of the process using Eq. (9.10):

$$U = aT_1^4 V_1.$$ (9.48)

There are the basic thermodynamic rules for equilibrium radiation. It should be stressed that the thermodynamic system treated in this chapter, "equilibrium radiation in a closed cavity," is a "simple" system whose only form of work is work of expansion. The differences in the thermodynamic description of this system from the usual "simple" systems are due only to the specific properties of the equation of state for the photon gas.

CHAPTER 10

Elasticity of Solids

10.1. Basic Thermodynamic Relations for Solids

It is known that if a solid (for example, a rod) is under a tensile (or compressive) force then the dimension of the solid will increase or decrease in the direction in which the force acts, i.e., the solid is deformed. We recall that the deformation of a solid can be permanent (i.e., such that the deformation caused by the force does not vanish after removal of the force) or elastic (i.e., such that the deformation does vanish after removal of the external force).

The greatest practical interest lies in study of elastic deformation. By elastic we mean the property of the body of returning to its shape and volume (for solids) or only volume (for gases and liquids) after removal of the external forces. It is apparent that thermodynamic methods apply. As for the residual (or, as it is sometimes called, plastic) deformation, it is essentially an irreversible process and ordinary thermodynamic methods do not apply to it.

Below we will describe tension (compression) in an elastic rod.†

We will denote the force acting on the solid (rod) by Ψ, and the

---

†In this chapter we always have in mind an isotropic solid, which is quite important for the subsequent treatment. A more general treatment can be found in Landau and Lifshits, Theory of Elasticity, Izd. "Nauka" (1965).

length of the rod along the direction in which the force acts
(the rod axis) by l. It is clear that Ψ is the generalized force and
l the generalized coordinate for a rod. For an elastic deformation
of a solid due to an external force (the so-called external load) the
combined equation of the first and second laws of thermodynamics
is written in the form

$$T\,dS = dU + p\,dV - \Psi\,dl \qquad\qquad (10.1)$$

in agreement with (1.28a) (the minus sign in front of the term
which expresses the elementary work of the elastic force as the
solid gets longer indicates that in tension work is done on the solid).

In all cases considered here we will assume that the pressure p of the medium in which the rod is located remains constant.

Furthermore, as we see in what follows, as the rod gets
longer its volume as a rule changes only slightly and for most
practically important cases we can assume with good accuracy
that V = const, i.e., dV = 0. For these cases, Eq. (10.1) assumes
the form

$$T\,dS = dU - \Psi\,dl. \qquad\qquad (10.2)$$

We also consider the case where V $\neq$ const.

According to (2.144a), (2.147a), (2.150a), and (2.153a) for
a system in which elastic forces act the Maxwell equations (for
the case where either p = const or V = const) will be

$$\left(\frac{\partial l}{\partial T}\right)_{S,V} = \left(\frac{\partial S}{\partial \Psi}\right)_{l,V}, \qquad\qquad (10.3)$$

$$\left(\frac{\partial l}{\partial S}\right)_{\Psi,p} = -\left(\frac{\partial T}{\partial \Psi}\right)_{S,p}, \qquad\qquad (10.4)$$

$$\left(\frac{\partial l}{\partial S}\right)_{T,V} = -\left(\frac{\partial T}{\partial \Psi}\right)_{l,V}, \qquad\qquad (10.5)$$

$$\left(\frac{\partial l}{\partial T}\right)_{\Psi,p} = \left(\frac{\partial S}{\partial \Psi}\right)_{l,V}. \qquad\qquad (10.6)$$

It is further evident that the magnitude of the tensile (compressive) force Ψ acting on the rod can be expressed as

$$\Psi = \psi\Omega, \qquad\qquad (10.7)$$

where Ω is the cross-sectional area of the rod and ψ is the tensile force per unit transverse area of the rod (the so-called stress).

In practice the change in the dimensions of a solid during deformation is expressed in terms of the relative length change ε, defined by

$$\varepsilon = \frac{l - l_0}{l}, \tag{10.8}$$

where l_0 is the length of the rod in the absence of a load and l is the length of the rod with a load Ψ. It is evident from this that

$$\varepsilon = 1 - \frac{l_0}{l} \tag{10.9}$$

and

$$d\varepsilon = \frac{l_0 dl}{l^2} \approx \frac{dl}{l}, \tag{10.10}$$

i.e.,

$$dl = l \, d\varepsilon. \tag{10.11}$$

In view of (10.7) and (10.11), Eq. (10.2) can be represented in the following form:

$$T \, dS = dU - \psi \Omega l_0 \, d\varepsilon. \tag{10.12}$$

The product Ωl_0 is the volume V of the rod.[†] Consequently,

$$T dS = dU - \psi V \, d\varepsilon. \tag{10.13}$$

If the volume V of the rod does not change during extension then Eq. (10.13) can be transformed in the following way[‡]

$$T ds_v = du_v - \psi \, d\varepsilon, \tag{10.14}$$

[†]Strictly speaking we must distinguish between the volume of a solid before and after deformation. Below all thermodynamic quantities will be referred to unit volume of the undeformed material.
[‡]Since $S = s_v V$, $dS = V ds_v + s_v dV$, and since $V = $ const, $dV = 0$ and thus $dS = V ds_v$; similarly, $dU = V du_v$.

where s_v and u_v are the bulk entropy and internal energy of the rod (whose dimensions are, respectively, $kcal/m^3 \cdot deg$ and $kcal/m^3$).

It is evident that the Maxwell equations for the elastically deformed rod can also be written in the variables s_v, u_v, ψ, and ε. It is not hard to show that

$$\left(\frac{\partial \varepsilon}{\partial T}\right)_{s_v, V} = \left(\frac{\partial s_v}{\partial \psi}\right)_{\varepsilon, V},$$

(10.15)

$$\left(\frac{\partial \varepsilon}{\partial s_v}\right)_{\psi, p} = -\left(\frac{\partial T}{\partial \psi}\right)_{s_v, p},$$

(10.16)

$$\left(\frac{\partial \varepsilon}{\partial s_v}\right)_{T, V} = -\left(\frac{\partial T}{\partial \psi}\right)_{\varepsilon, V},$$

(10.17)

$$\left(\frac{\partial \varepsilon}{\partial T}\right)_{\psi, p} = \left(\frac{\partial s_v}{\partial \psi}\right)_{T, p}.$$

(10.18)

10.2. The Equation of State for an Elastically Deformed Rod

We now consider the equation of state for an elastically deformed rod. For the case we are analyzing where $p = const$ and $V = const$ the equation of state in general is given by a relation of the form

$$l = f(T, \psi).$$

(10.19)

The equation of state for an elastically deformed rod widely used in elasticity theory is known as Hooke's law formulated in (1660), and is usually written as

$$\psi = E\varepsilon.$$

(10.20)

Here ε is the relative deformation of the rod and E is a proportionality coefficient (the so-called elastic stress modulus or Young's modulus). In agreement with Hooke's law the relative deformation of a rod is linearly related to the amount of stress acting on the rod.

Hooke's law is an approximate equation of state for an elastically deformed rod. Indeed, it is clear from (10.19) that l depends only on the stress ψ along an isotherm $T = const$. We assume now that l depends linearly on ψ along an isotherm:

$$l = l_0 + \left(\frac{\partial l}{\partial \psi}\right)_T \psi,$$

(10.21)

where $(\partial l/\partial \psi)_T$ is the magnitude of the constant for a given isotherm. Relation (10.21) can be transformed as follows:

$$\frac{l-l_0}{l} = \frac{1}{l}\left(\frac{\partial l}{\partial \psi}\right)_T \psi.$$ (10.22)

In view of (10.8) it is clear that Eq. (10.22) is identical to the equation for Hooke's law (10.20); here the elastic modulus E is defined as follows:

$$E = l\left(\frac{\partial \psi}{\partial l}\right)_T.$$ (10.23)

In general the relation $l(\psi)$ along an isotherm ultimately has a more complex character. However, with a precision which is quite sufficient for the overwhelming majority of applications, Hooke's law, which is the simplest equation for the isotherms of an elastically deformed rod, is valid. In other words, for the overwhelming majority of materials the Young's modulus E with T = const remains constant for any value of the elastic deformation ε. (Note that the linear relation can only be valid for small deformations. However, since only small deformations are elastic for most materials, Eq. (10.20) turns out to be quite valid for any elastic deformation; consequently there is no need to use a more complex power law.) In addition it should be noted that for certain materials such as stone, concrete, cast iron, and especially a number of plastics, E varies significantly with a change in ε. In what follows, however, we will assume that E is independent of ε.

The elastic modulus E is ultimately different for different materials. E is an individual characteristic of the elastic properties of the given material. It is evident from (10.20) that the greater E, the less the rod is deformed for the same stress. For different materials E varies over broad limits. At room temperature for many metals $E \approx 1 \times 10^6$ kg/cm^2; for granite $E = 0.49 \times 10^6$ kg/cm^2, for concrete $E \approx 0.2 \times 10^6$ kg/cm^2, while for rubber E is only 0.00008×10^6 kg/cm^2.

For some materials the elastic modulus is different for extension and compression. However, this difference is usually neglected in technical calculations.

The elastic modulus changes with changing temperature; E decreases with increasing temperature; consequently the greater the temperature the easier the material is to deform. Figure 10.1 shows the temperature dependence of the elastic modulus for steel.

For most materials the temperature dependence of E is quite weak between −50° and 50°C; thus for example for steel, as the temperature varies within these limits (−50° to 50°C), E changes by only about 2%. At these temperatures dE/dT = −438 kg/cm$^2 \cdot$ deg for steel. In many cases in carrying out practical

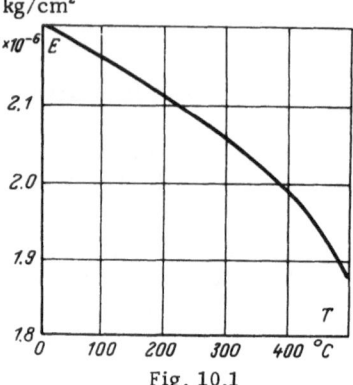

Fig. 10.1

calculations this relation can be neglected. For more precise calculations we can limit ourselves to a linear relation between E and T,

$$E = E_0 + \frac{dE}{dT}(T - T_0), \qquad (10.24)$$

where E_0 is the value of E at 0°C (it has previously been noted that we assume E to be independent of ε; consequently E depends only on the temperature and therefore Eq. (10.24) contains not the partial but the total derivative dE/dT).

As for the temperature dependence of the length of the rod, as found experimentally for fairly low temperatures, it can be assumed linear to a good approximation:

$$l = l_0(1 + \alpha_l t), \qquad (10.25)$$

where α_l is the linear thermal expansion coefficient of the material in the rod, by definition equal to

$$\alpha_l = \frac{1}{l}\left(\frac{\partial l}{\partial T}\right)_p, \qquad (10.26)$$

and t = $T - T_0$.

It is not hard to establish the relation between α_l and the bulk thermal expansion coefficient ordinarily used in thermodynamics, defined as

$$\alpha_v = \frac{1}{v}\left(\frac{\partial v}{\partial T}\right)_p. \qquad (10.27)$$

It is apparent that for not too large a temperature difference $t = T - T_0$, we can, with a good degree of precision, write

$$l_T = l_0 + \left(\frac{\partial l}{\partial T}\right)_p t \tag{10.28}$$

and

$$V_T = V_0 + \left(\frac{\partial V}{\partial T}\right)_p t . \tag{10.29}$$

Here $(\partial l/\partial T)_p$ and $(\partial V/\partial T)_p$ are values averaged over the temperature range from T_0 to T.

In view of (10 26) and (10.27) these relations can be represented in the form

$$l_T = l_0 (1 + \alpha_l t) \tag{10.30}$$

and

$$V_T = V_0 (1 + \alpha_v t). \tag{10.31}$$

We now consider a cube with edge l_0 at temperature T_0. The volume of the cube is

$$V_0 = l_0^3. \tag{10.32}$$

The volume of the same cube at the temperature T will obviously be

$$V_T = l_T^3 \tag{10.33}$$

or, in view of (10.30) and (10.32),

$$V_T = V_0 (1 + \alpha_l t)^3, \tag{10.34}$$

from which we find that

$$V_T = V_0 (1 + 3\alpha_l t + 3\alpha_l^2 t^2 + \alpha_l^3 t^3). \tag{10.35}$$

Since α_l is small relative to unity, neglecting the terms containing α_l^2 and α_l^3 because of their smallness we can with good precision write this relation in the form

$$V_T = V_0 (1 + 3\alpha_l t). \tag{10.36}$$

Using (10.31), we find

$$\alpha_v = 3\alpha_l, \tag{10.37}$$

i.e., the bulk thermal expansion coefficient is three times the linear coefficient.[t]

Knowing the relation between l and ψ along an isotherm (Hooke's law) and having data on the thermal expansion coefficient of the material in the rod, we can construct the state diagram of an elastically deformed rod. Figure 10.2 shows the $l-\psi$ diagram

---

[t]It should be borne in mind that this is only valid for isotropic solids.

of such a rod. As seen from this diagram, the isotherms are
linear in agreement with Hooke's law. The greater the temperature
the higher the isotherms lie in the $l - \psi$ diagram (since a higher
temperature corresponds to a longer rod because of thermal expansion). There is an increase in the slopes of the isotherms
with increasing temperature; this is due to the decrease in the
elastic modulus E with increasing temperature.

As we see, it is possible to calculate the thermal properties
of a stressed rod without particular difficulty.

10.3. The Caloric Properties of an Elastically Deformed Rod

We now turn to an analysis of the effect of elastic deformation
on the caloric properties of a rod.

The volume-specific entropy s_v as a function of the relative
deformation ε for an elastically deformed rod at constant temperature T is given by the Maxwell equation (10.17), while the
Maxwell equation (10.18) gives it as a function of the stress ψ.

Using (10.17)

$$\left(\frac{\partial s_v}{\partial \varepsilon}\right)_{T,V} = -\left(\frac{\partial \psi}{\partial T}\right)_{\varepsilon,V}$$

to calculate s_v, it is evident that

$$s_v(T, \varepsilon) = s_v(T, \varepsilon = 0) + \int_0^\varepsilon \left(\frac{\partial s_v}{\partial \varepsilon}\right)_{T,V} d\varepsilon. \tag{10.38}$$

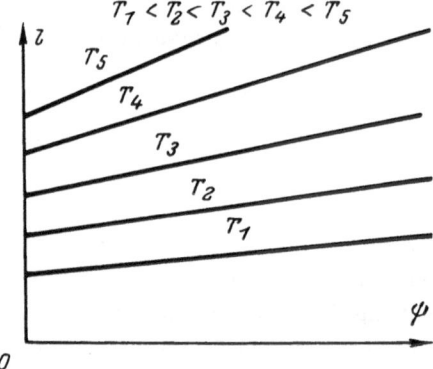

Fig. 10.2

Here $s_v(T, \varepsilon = 0)$ is the volume-specific entropy of the unstressed (underformed, $\varepsilon = 0$) rod at a temperature T, while $s_v(T, \varepsilon)$ is the volume-specific entropy of the rod at the same temperature T but with a relative deformation ε (remember that according to previous discussion the volume V of the rod does not change during deformation). In agreement with Eq. (10.17) we transform (10.38) to the following form:

$$s_v(T, \varepsilon) = s_v(T, \varepsilon = 0) - \int_0^\varepsilon \left(\frac{\partial \psi}{\partial T}\right)_{\varepsilon, V} d\varepsilon. \tag{10.39}$$

A knowledge of the state of an elastically deformed rod — Hooke's law — allows us to carry out the integration on the right-hand side of Eq. (10.39). In fact, we find from Eq. (10.20) that

$$\left(\frac{\partial \psi}{\partial T}\right)_{\varepsilon, V} = \frac{dE}{dT} \varepsilon. \tag{10.40}$$

Since the elastic modulus E (and consequently also dE/dT) does not depend on ε, dE/dT can be brought outside the integral sign during the integration, and we find from (10.39) that

$$s_v(T, \varepsilon) = s_v(T, \varepsilon = 0) - \frac{dE}{dT} \frac{\varepsilon^2}{2}. \tag{10.41}$$

Equation (10.41) gives the entropy s_v of a rod along an isotherm as a function of the relative deformation ε. From this equation it is not hard to find the dependence of s_v on the stress ψ; replacing ε in (10.41) using (10.20) and assuming that there is no deformation ($\varepsilon = 0$) which corresponds to no stress ($\psi = 0$), we find

$$s_v(T, \psi) = s_v(T, \psi = 0) - \frac{dE}{dT} \frac{\psi^2}{2E^2}. \tag{10.42}$$

As we see from this equation, the entropy of a rod always increases as it is deformed; since we always have $dE/dT < 0$ while ψ is squared, for either tension or compression we have $s_v(T, \psi) > s_v(T, \psi = 0)$.

Equation (10.42) can also be obtained in a different way from that given above; from the obvious relation

$$s_v(T, \psi) = s_v(T, \psi = 0) + \int_0^\psi \left(\frac{\partial s_v}{\partial \psi}\right)_{T, p} d\psi \tag{10.43}$$

using Eq. (10.18)

$$\left(\frac{\partial s_v}{\partial \psi}\right)_{T, p} = \left(\frac{\partial \varepsilon}{\partial T}\right)_{\psi, p},$$

and the equation for Hooke's law (10.20), we can easily obtain Eq. (10.42).

We can calculate the other caloric properties of an elastically deformed rod in a similar way.

In fact, all of these calculations are arbitrary to the extent of the caloric properties of the rod in the absence of deformation ($\varepsilon = 0$, $\psi = 0$). These "zero" functions (which here play the role that for a real gas is played by the caloric properties of an ideal gas) cannot be calculated by thermodynamic methods alone.

We find for the volume-specific internal energy u_V of the rod

$$u_v(T, \varepsilon) = u_v(T, \varepsilon = 0) + \int_0^\varepsilon \left(\frac{\partial u_v}{\partial \varepsilon}\right)_{T,V} d\varepsilon, \tag{10.44}$$

and it is evident from Eq. (10.14) that

$$\left(\frac{\partial u_v}{\partial \varepsilon}\right)_{T,V} = \psi + T\left(\frac{\partial s_v}{\partial \varepsilon}\right)_{T,V}, \tag{10.45}$$

whence, using (10.17), we find

$$\left(\frac{\partial u_v}{\partial \varepsilon}\right)_{T,V} = \psi - T\left(\frac{\partial \psi}{\partial T}\right)_{\varepsilon,V}. \tag{10.46}$$

Using Eqs. (10.46) and (10.20), we find from (10.44)

$$u_v(T, \varepsilon) = u_v(T, \varepsilon = 0) + \left(E - T\frac{dE}{dT}\right)\frac{\varepsilon^2}{2} \tag{10.47}$$

or, what is the same,

$$u_v(T, \psi) = u_v(T, \psi = 0) + \left(1 - \frac{T}{E}\frac{dE}{dT}\right)\frac{\psi^2}{2E}. \tag{10.48}$$

The enthalpy of an elastically deformed rod is calculated
as follows. In agreement with Eq. (2.38) the total enthalpy of the
whole rod is written as†

$$I^* = U + pV - \Psi l. \cdot \tag{10.49}$$

Using (10.7), we find that

$$I^* = U + pV - \psi \Omega l, \tag{10.50}$$

while since

$$\Omega l = V,$$

dividing both sides of Eq. (10.50) by V, we find

$$i^*_v = u_v + p - \psi. \tag{10.51}$$

Here i^*_v and u_v are, respectively, the bulk enthalpy and the internal
energy of the rod.

Thus the change in the enthalpy of the rod with deformation
for T = const (as previously in the assumption that the volume
V of the rod and the pressure p of the surrounding medium do
not change) can be represented simply in the form

$$i^*_v(T, \psi) = i_v(T) - \psi \tag{10.52}$$

or, what is the same thing,

$$i^*_v(T, \varepsilon) = i_v(T) - E\varepsilon. \tag{10.53}$$

Knowing s_v, u_v, and i^*_v it is easy to calculate the volume-
specific free energy $f_v = u_v - Ts_v$ and the isobaric−isothermal
potential $\varphi^*_v = i^*_v - Ts_v$.

These are the relations which determine the changes in the
caloric properties of an elastic rod when it is deformed.

---

†Remember that in agreement with Section 2.1, the asterisk indicates the enthalpy
for the present system (which differs from the "usual" enthalpy defined by I = U + pV).

We now consider the thermal diagrams for an elastically deformed rod (s, ψ; u, ψ; i, ψ).

Everywhere above we have calculated volume-specific values of the caloric properties. This was only done to simplify the calculations and not for any other reason. Since it is more common to operate with weight-specific values we recall that we can easily transform to weight-specific values if we know the specific gravity (or specific volume) of the material:

$$s = \frac{s_v}{\gamma}, \quad u = \frac{u_v}{\gamma}, \quad i = \frac{i_v}{\gamma}. \tag{10.54}$$

Figure 10.3 shows the s–ψ diagram for an elastic rod. In agreement with Eq. (10.42) the isotherms in this diagram are parabolas. The greater the temperature corresponding to a given isotherm the higher lies the isotherm. The diagram uses the weight-specific entropy s = s_v/γ. Since E and γ decrease with increasing temperature while dE/dT increases in absolute value, it is clear from (10.42) that the higher the temperature the steeper the parabola representing the isotherm.

The u–ψ diagram of an elastic rod has a similar character, as is evident from an analysis of Eq. (10.48).

The enthalpy-stress diagram is shown in Fig. 10.4. Figure 10.4a shows the i_v^*–ψ diagram. According to Eq. (10.52) i_v^* decreases linearly with increasing ψ, and the isotherms in this diagram are lines parallel to each other; it follows from (10.52) that the tangent of the slope of an isotherm is $(\partial i_v^*/\partial \psi)_T = -1$. The diagram for the weight-specific enthalpy $i^* = i_v^*/\gamma$ is shown in Figure 10.4b. The difference between this diagram and the preceding one is that since, as is not hard to establish, $(\partial i^*/\partial \psi)_T = -1/\gamma$,

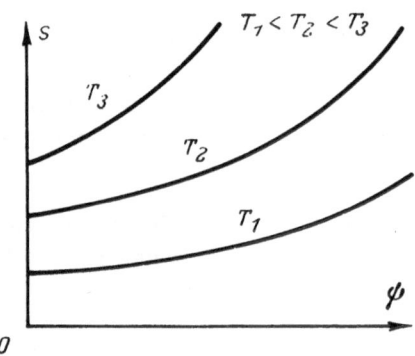

Fig. 10.3

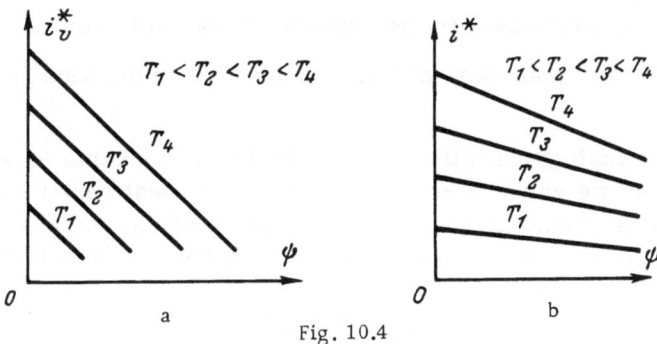

Fig. 10.4

the isotherms are already not parallel to each other; since as is well known the specific gravity of a material decreases with increasing temperature, the higher the temperature the steeper the slope of the isotherm.

Such are some of the thermal diagrams of state for an elastically deformed rod.

We now discuss some important changes in the internal energy and enthalpy due to deformation of the rod. Calculations of these quantities done using Eqs. (10.48) and (10.52) for a steel rod at room temperature in which there is a tensile stress ψ = 20 kg/mm^2 give the following results:

The change in internal energy is

$$u_v(T, \psi) - u_v(T, \psi=0) = 0.023 \text{ cal/cm}^3,$$

or in weight units (γ = 8 g/cm^3)

$$u(T, \psi) - u(T, \psi = 0) = 0.0028 \text{ cal/g},$$

and the change in enthalpy is

$$i^*_v(T, \psi) - i_v(T) = -46.8 \text{ cal/cm}^3,$$

or in weight units

$$i^*_v(T, \psi) - i(T) = -5.86 \text{ cal/g}.$$

As we see from this example the change in the internal energy during isothermal extension of a rod is quite small while the change (decrease) in the enthalpy of the rod is quite substantial.

We now consider the heat capacity of an elastically deformed rod.

By analogy with the isobaric (c_p) and isochoric (c_v) heat capacities we can introduce the concepts of heat capacities of a material at constant stress c_ψ (the analog of c_p) and at constant deformation c_ε (the analog of c_v). In agreement with (1.37) and (1.38) it is clear that

$$c_\psi = T \left(\frac{\partial s}{\partial T} \right)_\psi \tag{10.55}$$

and

$$c_\varepsilon = T \left(\frac{\partial s}{\partial T} \right)_\varepsilon . \tag{10.56}$$

Using Eq. (10.41) and differentiating with respect to T once with ψ = const and another time with ε = const and assuming that in both cases the pressure of the medium does not change (p = const), we find

$$c_{\psi,p}(T,\psi) = c_{\psi,p}(T, \psi = 0) - \frac{T\varepsilon^2}{2\gamma} \frac{d^2E}{dT^2} + \frac{T\psi^2}{\gamma E^3} \left(\frac{dE}{dT} \right)^2 \tag{10.57}$$

and

$$c_{\varepsilon,p}(T,\varepsilon) = c_{\varepsilon,p}(T, \varepsilon = 0) - \frac{T\varepsilon^2}{2\gamma} \frac{d^2E}{dT^2} \tag{10.58}$$

(here the heat capacity is already given not in volume-specific but in weight-specific units).

It is to be understood that the heat capacities $c_{\varepsilon, p}$ and $c_{\psi, p}$ are simply equal to c_p in the absence of deformation, i.e.,

$$c_{\varepsilon,p}(T, \varepsilon = 0) = c_{\psi,p}(T, \psi = 0) = c_p. \tag{10.59}$$

It follows from (10.57) and (10.58) using (10.59) that

$$c_\psi - c_\varepsilon = \frac{T\psi^2}{\gamma E^3} \left(\frac{dE}{dT} \right)^2 . \tag{10.60}$$

This result can be found in another way using equation (2.156) obtained in Chapter 2

$$c_\xi - c_x = T \left(\frac{\partial \xi}{\partial T} \right)_x \left(\frac{\partial x}{\partial T} \right)_\xi , \tag{10.61}$$

which can be represented in the form

$$c_\xi - c_x = \frac{T}{\gamma V}\left(\frac{\partial \xi}{\partial T}\right)_X \left(\frac{\partial X}{\partial T}\right)_\xi .$$
(10.62)

Since $\xi = -\Psi$ and $X = l$ in the case of interest, we have

$$c_\Psi - c_l = -\frac{T}{\gamma V}\left(\frac{\partial \psi}{\partial T}\right)_l \left(\frac{\partial l}{\partial T}\right)_\Psi ,$$
(10.63)

whence, using (10.7), (10.11), and (10.20), we find (10.60).

As shown by calculations, the quantities

$$T\varepsilon^2 (d^2 E/dT^2)/2\gamma \quad \text{and} \quad T\psi^2 (dE/dT)^2/\gamma E^3$$

which enter Eqs. (10.57) and (10.58) in most cases are negligible relative to c_p. Therefore to a high degree of precision we can assume

$$c_{\varepsilon,p} = c_{\psi,p} = c_p,$$
(10.64)

for any values of ε and ψ, i.e., the elastic deformation does not exert significant influence on the heat capacity of the material in the rod.

10. 4. Adiabatic and Isothermal Deformation of a Rod

We consider the changes in the temperature of an elastic rod due to its deformation.

We assume that the extension of the rod occurs fast enough that the elastic deformation process can be assumed adiabatic (as usual we assume that the volume of the rod does not change during deformation and the pressure of the surrounding medium is constant).

The change in the temperature of the rod during adiabatic extension can be calculated from the usual equation

$$T = T_0 + \int_0^\phi \left(\frac{\partial T}{\partial \psi}\right)_{\varepsilon,p} d\psi,$$
(10.65)

where T_0 and T are the temperatures of the rod, respectively, in the undeformed and deformed states.

To calculate $(\partial T/\partial \psi)_{s, p}$ we use the Maxwell equation (10.16)

$$\left(\frac{\partial T}{\partial \psi}\right)_{s_v, p} = -\left(\frac{\partial \varepsilon}{\partial s_v}\right)_{\psi, p} .$$

The derivative $(\partial \varepsilon/\partial s_v)_{\psi, p}$ is transformed as follows:

$$\left(\frac{\partial \varepsilon}{\partial s_v}\right)_{\psi, p} = \left(\frac{\partial \varepsilon}{\partial T}\right)_{\psi, p} \left(\frac{\partial T}{\partial s_v}\right)_{\psi, p} . \tag{10.66}$$

In turn,

$$\left(\frac{\partial T}{\partial s_v}\right)_{\psi, p} = \frac{T}{c_{\psi, p} \gamma} \tag{10.67}$$

or, in agreement with (10.64), simply

$$\left(\frac{\partial T}{\partial s_v}\right)_{\psi, p} = \frac{T}{c_p \gamma} \tag{10.68}$$

(here $c_{\psi, p} = c_p$ is the weight–specific heat capacity).

Using Eq. (10.11) we can write $(\partial \varepsilon/\partial T)_{\psi, p}$ as follows:

$$\left(\frac{\partial \varepsilon}{\partial T}\right)_{\psi, p} = \frac{1}{l} \left(\frac{\partial l}{\partial T}\right)_{\psi, p} . \tag{10.69}$$

The quantity on the right–hand side of this equation is none other than the linear thermal expansion coefficient for the material in the rod, α_l (10.26). Consequently,

$$\left(\frac{\partial \varepsilon}{\partial T}\right)_{\psi, p} = \alpha_l, \tag{10.70}$$

and Eq. (10.65) can be written in the following form:

$$T = T_0 - \int_0^\psi \frac{\alpha_l T}{c_p \gamma} d\psi. \tag{10.71}$$

Assuming that, like the elastic modulus, α_l is nearly independent of ψ, we find from Eq. (10.71) that

$$T = T_0 - \frac{\alpha_l T_{av} \psi}{c_p \gamma} ; \tag{10.72}$$

for a comparatively small difference $T-T_0$, where $T_{av} = (T + T_0)/2$ and α_l, c_p, and γ are averaged over the temperature range from T_0 to T.

The equation which relates T to T_0 can be obtained in a different form. It is evident from (10.16) that with $s = \text{const}$ we have

$$dT = -\left(\frac{\partial \varepsilon}{\partial s_v}\right)_{\psi,\,p} d\psi \tag{10.73}$$

or, using (10.68) and (10.70),

$$\frac{dT}{T} = -\frac{\alpha_l}{c_p\gamma}\, d\psi. \tag{10.74}$$

Integrating this relation, we find

$$\ln\frac{T}{T_0} = -\frac{\alpha_l\psi}{c_p\gamma}, \tag{10.75}$$

so that

$$T = T_0 e^{-\frac{\alpha_l\psi}{c_p\gamma}}. \tag{10.76}$$

Expanding the exponential in series and retaining only the first two terms (which is admissible due to the smallness of $T-T_0$), we find

$$T = T_0 - \frac{\alpha_l T_0 \psi}{c_p\gamma},$$

which agrees with (10.72).

We consider this important equation in more detail. It shows that the amount of temperature change when the rod is extended does not depend on the length nor on the cross-sectional area of the rod. Since T, c_p, and γ are positive, the sign of the difference (T_0-T) is determined by the signs of ψ and α_l. The value of ψ is positive if the rod is extended and negative if it is compressed. As for the thermal expansion coefficient α_l, for the overwhelming majority of solids $\alpha_l > 0$ so that during extension $T < T_0$ and the rod is cooled, while during compression $T > T_0$ and the rod is heated. In rare cases (for example, for rubber which has not been too strongly extended), we have $\alpha_l < 0$ and the sign of the effect changes (the rubber gets hotter as it is extended).

As shown by calculations and experiments, the amount a rod is heated during adiabatic extension (compression) is quite perceptible. Thus during extension of a steel wire with a stress $\psi = 20$ kg/mm^2

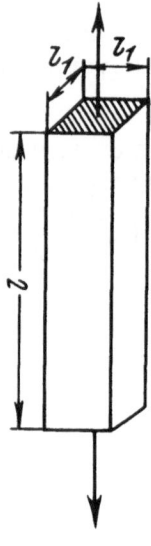

Fig. 10.5

the temperature of the wire drops by 0.16°C. This value can easily
be measured experimentally with good precision. It is interesting
to note that in the 1880s experiments on adiabatic extension of silver
and steel wires were used to calculate the mechanical equivalent
of heat.

In principle the adiabatic extension of a wire can be used
for an experimental determination of the heat capacity c_p of the
wire material.

If a rod is elastically deformed under isothermal conditions
then the rod can exchange heat with the surroundings, and it is
clear that heat will be absorbed during extension of the rod while
during compression heat will flow out into the surrounding medium.
We will calculate the amount of heat absorbed (or liberated) by the
rod during isothermal deformation.

The change in the entropy of a rod during isothermal deformation is given by the obvious relation

$$s_{v2}(T, \psi) - s_{v1}(T, \psi = 0) = \int_0^\psi \left(\frac{\partial s}{\partial \psi} \right)_{T,p} d\psi. \qquad (10.77)$$

Using the Maxwell equation (10.18)

$$\left(\frac{\partial s_v}{\partial \psi}\right)_{T,p} = \left(\frac{\partial \varepsilon}{\partial T}\right)_{\psi,p}$$

and Eq. (10.70), we find

$$s_{v_2}(T,\psi) - s_{v_1}(T,\psi=0) = a_l\psi, \tag{10.78}$$

and correspondingly per unit weight of the rod,

$$s_2 - s_1 = \frac{a_l\psi}{\gamma}. \tag{10.79}$$

Since, as is well known, in an isothermal process we have

$$q_{1-2} = T(s_2 - s_1),$$

we find, using (10.79), that

$$q_{1-2} = \frac{Ta_l\psi}{\gamma}. \tag{10.80}$$

As shown by calculations, q_{1-2} is comparatively small. Thus, for example, for a steel rod with a stress of 20 kg/mm^2 q_{1-2} is 0.02 kcal/kg.

These are the basic features of the thermodynamic description of the elastically deformed rod, assuming that the volume of the rod does not change during extension (or compression).

We now consider the change in the volume of a solid during extension. We will consider a rod of length l whose transverse cross section is a square of side l_1 (Fig. 10.5). As shown by experiment, during extension (compression) of the rod its volume does change somewhat: with an increase in its length by Δl each side of its cross section decreases in length by Δl_1. We will assume a relative longitudinal deformation of

$$\varepsilon = \frac{\Delta l}{l}, \tag{10.81}$$

and a relative transverse deformation of

$$\varepsilon_1 = \frac{\Delta l_1}{l_1}. \tag{10.82}$$

The quantity

$$\mu = \frac{\varepsilon_1}{\varepsilon} \tag{10.83}$$

is called the transverse deformation coefficient (Poisson coefficient). It is found experimentally that for many materials $\mu = 0.25$ to 0.3.

If l and l_1 are, respectively, the length and the width of the rod before deformation then the volume of the rod before deformation is $V_0 = l l_1^2$. The length of the
deformed rod is $(l + \Delta l)$, or $l(1 + \varepsilon)$ using (10.81). The width of the deformed rod is
$(l_1 - \Delta l)$ or, using (10.82) and (10.83),

$$l_1 (1 - \varepsilon_1) = l_1 (1 - \mu\varepsilon). \tag{10.84}$$

Thus the cross-sectional area of the deformed rod will be $l_1^2(1-\mu\varepsilon)^2$ and its volume
will be

$$V = l (1 + \varepsilon) \, l_1^2 (1 - \mu\varepsilon)^2 = V_0 (1 + \varepsilon) (1 - \mu\varepsilon)^2. \tag{10.85}$$

Multiplying the expressions in parentheses and assuming that since ε is small we can
neglect the terms containing ε^2 and ε^3, we find

$$V = V_0 [1 + \varepsilon (1 - 2\mu)] \tag{10.86}$$

and, further,

$$\frac{V - V_0}{V_0} = \varepsilon (1 - 2\mu). \tag{10.87}$$

It follows from this that

$$\frac{dV}{V_0} = (1 - 2\mu) \, d\varepsilon. \tag{10.88}$$

As we see from Eq. (10.87), with increasing stress ψ the volume of the rod increases more the smaller the values of μ and E. Evidently with $\mu = 0.5$ the volume
of a body does not change during deformation. An example of a material in which
$\mu = 0.5$ is rubber — its volume does not change during extension.

In a thermodynamic treatment of an elastically deformed rod the change in
the volume of the rod during extension (compression) can be taken account of as
follows. The combined equation for the first and second laws of thermodynamics for
an elastically deformed solid (10.1) is

$$T \, dS = dU + p \, dV - \Psi \, dl$$

and, using (10.88), (10.7), and (10.11), this can be transformed to

$$T dS = dU + V_0 [(1 - 2\mu) \, p - \psi] \, d\varepsilon \tag{10.89}$$

or

$$T ds_v = du_v + [(1 - 2\mu) \, p - \psi] \, d\varepsilon. \tag{10.89a}$$

This equation is the analog of Eq. (10.14), which was used assuming that the
volume of the rod does not change during deformation. In contrast to Eq. (10.14), in

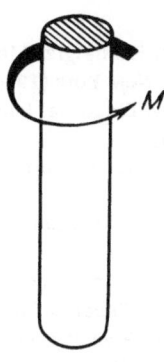

Fig. 10.6

Eq. (10.89a) the differential $d\varepsilon$ has the factor $[(1-2\mu)p-\phi]$ instead of ψ. Therefore, if we take account of changes in the volume of the rod during deformation, the calculations use not ψ but $[(1-2\mu)p-\psi]$. If the pressure p of the surrounding medium is small compared to the deforming stress ψ then considering the changes in the volume of the rod does not make any practical change in the calculations for V = const. If the pressure of the medium is commensurable with the stress ψ then changes in V will substantially affect the results of the calculation. Here $[(1-2\mu)p-\psi]$ can be quite small. Thus, for example, for the case of a steel rod (μ = 0.25) with a tensile stress of 5 kg/mm^2 within a high-pressure vessel in which a pressure of 1000 kg/cm^2 is maintained, $[(1-2\mu)p-\psi]$ is zero. This means that the work done deforming the rod is compensated by work of the external pressure forces which is equal in magnitude but opposite in sign. Here the caloric properties of the rod do not change, as if the rod were not deformed.

In conclusion we note that in addition to problems involving extension or compression of an elastic rod, in a number of cases it is of interest to consider an elastic rod which is subjected to a certain torsional moment M (Fig. 10.6).

The combined equation for the first and second laws of thermodynamics for this case is written in the form

$$T\,dS = dU + p\,dV - M\,d\omega, \qquad (10.90)$$

where ω is the angle through which the rod is twisted due to the moment M. The detailed thermodynamic analysis of this problem of torsion in a rod can be done as for the application above to an extended (compressed) rod.

Supplementary Readings

1. I. P. Bazarov, Thermodynamics, Fizmatgiz, Moscow, 1961; (translation) Pergamon, Oxford and MacMillan, New York (1964).
2. K. P. Belov, Elastic Thermal and Electrical Phenomena in Ferromagnets, 2nd. edition, Gostekhteorizdat (1957).
3. V. A. Kirillin, V. V. Sychev, and A. E. Sheindlin, Technical Thermodynamics, Izd. "Energiya" (1968).
4. L. D. Landau and E. M. Lifschitz, Statistical Physics 2nd. edition, Izd. "Nauka" (1964) (translation) 2nd. edition, Addison-Wesley, Reading, Mass. (1969); (German translation), Akademie-Verlag, Berlin (1969).
5. L. D. Landau and E. M. Lifschitz, Theory of Elasticity, Izd. "Nauka" (1965); (translation) 2nd. edition, Addison-Wesley, Reading, Mass. (1971); (German translation) Akademie-Verlag, Berlin (1969).
6. L. D. Landau and E. M. Lifschitz, Electrodynamics of Continuous Media, Fizmatgiz, Moscow (1959); (translation) Pergamon, Oxford and New York (1960); (German translation) Akademie-Verlag, Berlin (1967).
7. M. A. Leontovich, Introduction to Thermodynamics, Gostekhizdat (1950); (German translation) Deutsch. Verlag der Wiss. (1953).
8. H. A. Lorentz, Abhandlungen uber Theoretisch Physik, Bd. 1, Leipzig and Berlin (1970); Lectures on Theoretical Physics (translation) Vol. 1, MacMillan & Co., London (1927); [Note: Vol. 2 was never published].
9. G. I. Skanavi, Physics of Dielectrics (for Weak Fields), Gostekhizdat (1949).
10. D. Schoenberg, Superconductivity, Cambridge University Press (1960).
11. P. S. Epstein, Textbook of Thermodynamics, 4th edition, Wiley, N. Y. (1947).
12. N. Davidson, Statistical Mechanics, McGraw-Hill, N. Y. (1962).
13. D. Ruelle, Statistical Mechanics, Benjamin, N. Y. (1969).
14. Ya. P. Terletskii, Statistical Physics, North-Holland Publ. Co., Amsterdam (1971).
15. R. Jancel, Foundations of Classical and Quantum Statistical Mechanics, Pergamon, New York (1969).
16. A. Munster, Statistical Thermodynamics, Vol. 1, Springer-Verlag, New York (1969).
17. Z. M. Galasiewicz, Superconductivity and Quantum Fields, Pergamon, New York (1970).
18. H. E. Stanley (ed.), Cooperative Phenomena near Phase Transitions – a Bibliography with Selected Readings, M.I.T. Press, Cambridge, Mass. (1973).
19. C. Domb and M. S. Green (eds.), Phase Transitions and Critical Phenomena, Vol. 1, Vol. 2, Academic Press, New York (1973).
20. W. D. Gregory, W. N. Mathews, Jr. and E. A. Edelsack (eds.), The Science and Technology of Superconductivity, 2 vols, Plenum, New York (1973).
21. A. J. Bard, Chemical Equilibrium, Harper and Row, Evanston (1966).
22. B. M. King, Phase Equilibrium in Mixtures, Pergamon, New York (1969).
23. R. Haase, and H. Schonert, Solid-Liquid Equilibrium, Pergamon, New York (1969).
24. D. Wagner, Introduction to the Theory of Magnetism, Pergamon, New York (1972).
25. T. M. Reed and K. E. Gubbins, Applied Statistical Mechanics, McGraw-Hill, New York (1973).